Bitesize

THOUGHTFUL FOOD AND FOOD FOR THOUGHT
THE FOOD OF THE FUTURE

The future is cultured, not slaughtered

Iris & Harry Efthymiou-Egleton

Copyright © 2020 Iris Efthymiou-Egleton

All right reserved. NOTICE OF RIGHTS: All rights reserved. Except as permitted under the United States Copyright Act of 1976, no part of this publication may be reproduced or distributed in any form or by any means or stored in a database or retrieval system, without the prior written permission of the author.

Disclaimer

The information and recommendations provided herein are stated to be truthful and for educational purposes only Regardless of the claims stated in this book, the author and the publishers hold no responsibility for the outcomes of practicing the pieces of advice presented throughout the book. No warranties or guarantees are expressed or implied by the publisher's choice to include any of the content in this book. The content of all chapters is the sole expression and opinion of the authors and not necessarily that of the publisher. Neither the publisher nor the authors shall be liable for any physical, psychological, emotional, financial, or commercial damages, including, but not limited to, special, incidental, consequential, or other damages. The authors have strived to be as accurate and complete as possible in the creation of this book. While all attempts have been made to verify information provided in this publication, the authors assumes no responsibility for errors, omissions, or contrary interpretation of the subject matter herein. In practical advice books, like anything else in life, there are no guarantees of results. Readers are cautioned to rely on their own judgment about their individual circumstances and to act accordingly. This book is an educational guide that provides general information about life and the future. The materials are "as is" and without warranties of any kind, either expresses or implied. This book is designed to provide information and motivation to the readers and not to replace any type of academic, psychological, financial, or other kinds of professional advice. As also mentioned elsewhere in the book, everyone is responsible for their own choices, actions, and results... Further research should be carried out by readers who intend to use the information contained in this book.

TABLE OF CONTENTS

AN INTRODUCTION TO THE FUTURE 6
AN INTRODUCTION TO THE PRESENT 8
THE CURRENT CHALLENGES 15
    WORLD HUNGER 15
    DEMOGRAPHIC & DIETARY CHANGES 16
    LAND DETERIORATION 20
    FOOD PACKAGING 22
    FOOD LOSS 24
    HEALTH PROBLEMS 28
    OBESITY 29
    DIABETES 33
THE FUTURE AROUND FOOD 36
    VEGANISM 36
    AN INDIVIDUALIZED DNA-BASED DIET 38
    ECONOMY: THE SUGAR TAX 45
FOOD PRODUCTION TRENDS 50
    THE CASE OF SUSTAINABLE INTENSIFICATION 52
    LAND SPARING 53
    LIVESTOCK WELFARE 54
    TECHNOLOGY 55
    3D PRINTING 56
    HACKING PHOTOSYNTHESIS 57
    SHARING SERVICES 59
    UNDERGROUND FARMING 59
    VERTICAL FARMING 60
    DESERT AGRICULTURE 62
    PRECISION AGRICULTURE & NANOTECHNOLOGY 63

- SATELLITES AND MOBILE RADIO ANTENNAS 65
- AUTONOMOUS TRACTORS 65
- SMART SILOS 66
- DRONES 68
- GM FOODS 68
- BUSINESS EXAMPLES 71

FOOD TRENDS 75
- A TASTE OF UMAMI 76
- (G)ASTRONOMY 78
- BIOMIMETICS 80
- ALTERNATIVES TO MEAT AND FISH 81
- ALGAE 84
- INSECTS 85
- BUSINESSES WITH NEW FOOD TRENDS 89

THE FUTURE OF OUR EATING HABITS 92
- PERSONALIZED NUTRITION: A STEP TOWARDS WELLNESS 97
- THE FUTURE OF FOOD DELIVERY 100
- AGGREGATORS: THE NEW DELIVERY 102
- MEET CLOUD KITCHENS 105
- THE FUTURE'S KITCHEN: A WELLNESS-CENTERED ROOM 106
- FOOD SHARING 110

THE FUTURE OF WATER 112

EPILOGUE 117

SOURCES 128

AN INTRODUCTION TO THE FUTURE

You wake up early in the morning. Rested from your deep yet natural sleep, you head to the bathroom where you are informed whether your sleeping patterns were optimal. Subsequently, while you're flossing and brushing your teeth, the mirror reminds you of your tasks, responsibilities, and meetings for the day.

Your personal medical bot informs you of any vitamin or mineral abnormalities that it detected through your personal waste and your presence in front of the mirror. With one approval click it sends an order to the kitchen computer to prepare your day's food so that the food itself compensates for any abnormalities s well as ensuring that you are getting the right amount of anti-aging medication.

All of this is feasible as the internet of things has become affordable for everyone. You will walk to your lunch meeting at a local restaurant and provide a swab to aid the restaurant in serving you a meal personalised to your nutritional needs. The meals are primarily super food-based and incredibly tasty, the seamless connections between you, your needs, wants and the reality of universal personalization has meant that this is now your new norm.

When you get home your dinner will not be prepared until you have informed the computer what you ate at your business lunch. The final meal of the day will contain all the elements that are needed to ensure that you have eaten exactly the right foods to give you the correct balance in your diet.

Working from home is now the custom not the exception, this paradigm shift will occur following the long term changes that will be brought about by the impact of a number of Covid19 type events on Government policies around the globe.

As a result of technological advances, National policies, and abundance, mundanity will be replaced by an optimization that has a profoundly positive effect on not just us, but on society as a whole. This optimization will become all-pervasive, stemming from the omniscient AI employed to facilitate our experiences and leading itself to the narcissus levels of individual focus lent to us within our restaurants, supermarkets, and homes.

AN INTRODUCTION TO THE PRESENT

Humankind is curious and controlling. Combine the two, and you can understand why we are so eager to know early on what our future holds. From the weather to the economy, we like to know as much as possible as early as possible. We also wish to keep evolving and revolutionizing; being promptly and adequately informed helps us do both.

Food is not exempt from our desire to learn about humanity's future prospects regarding the feasible and sustainable options that are constantly increasing in viability. Food is also not exempt from our evolutionary and revolutionary nature, either. From our cooking on the fire to cultivating fields and running farms, then from finding out a way to preserve food all the way to mass-producing and distributing it. And now, this century's food-related revolution is its digitization.

Yet, as impressive as all this is, big mistakes and grave failures were not uncommon. See, food is a highly complex subject; it includes what we eat, how much of it and how often we eat it, how we cook, produce and distribute it, as well as how we combine different foods. Eating is a daily need for survival, so it only makes sense to want to know where that food is coming from and how it impacts our health, our environment, our economy, our entire community, and culture.

Unfortunately, human ignorance, negligence, indifference and extreme lack of moderation has led to numerous problems, such as environmental pollution, extinction of animal and plant species as well as health issues ranging from nutrition-related ones, e.g., obesity and diabetes, both of which we will be examining in this book, to several epidemics or pandemics. All of these issues are the product of poor foresight and planning.

The factors that negatively impact food security are global. Following the current path will not solve the problem of hunger and the further problems created by it. Drastic change is needed, and the drastic changes that are starting to take place are the main focus of this book.

In terms of diet, what's needed on this planet in the coming years is a steadier, more secure and more responsible way of providing nutrition to the world in a way that takes into consideration its waste, the sustainability of its products and the recycling of water, all without taking taste and enjoyment away from food and the act of eating.

The world's **population has been increasing faster than food production**, even with modern agricultural technology. There will be nearly 10 billion people to feed by 2050. Two-thirds of these people will live in urban areas. One of the issues that scientists are going to have to deal with is how ten billion mouths can be fed. Most of the population growth is expected to occur in the urban regions of developing countries. 90 percent of this growth will be in Asia and Africa (UN 2017). At the same time, when it comes to food production in developing countries, there is a shift towards larger, more productive units, which is also accompanied by a rise in livestock numbers overall.

Worldwide, malnutrition affects billions of people. Plus, almost a billion people are undernourished. Undernourishment means that one's diet lacks enough calories and nutrients, causing health problems and compromising their cognitive development. And then there are even more people who don't eat right, including those who are "over-nourished" and/or those who have energy-dense, nutrient-poor diets that lead to chronic diseases.

Those numbers are truly horrific. According to Full Fact, more than one million Britons used food banks in 2014, while, according to The Independent, a new record was hit when nearly four million Britons accessed a food bank in 2017.

In the US, forty million Americans live in food poverty, a large proportion of them in need of food stamps.

So, the future of the food industry worldwide lies at the center of efforts, innovations, and initiatives to make more efficient use of limited resources to feed a growing global population. The more generic goal is to develop a restorative and regenerative circular economy that connects the cycles of food, water, and energy with human nutrition and health the best way possible. The more specific goal is to maximize recycling and the reuse of resources; in other words, to minimize waste.

Before saying anything more, I should clearly point out that the experts define the future of food focus on three health-based issues: Two of them being ailments, namely **diabetes and obesity**, and the third of which is none other than the devastating socioeconomic issue that is **hunger**.

Demographic changes, scarcity of natural resources, climate change, and an overwhelming amount of food waste are the four main global geographical and economic issues that the next agriculture models, the future models, must combat. In this age of such disruption, no actor can play on their own. Governments, investors, and innovators all need to come together and form a broad global sustainable collaboration with an aim to improve global food security, this could perhaps be achieved through agreements similar to the 2016 Paris climate accords.

So, yes, today's 'genetic' goals, in brief, are to relegate diabetes and obesity to health problems of the past while developing the most economical and most effective mechanisms to provide everyone with access to adequate fresh and healthy food products.

Food falls into the category of those goods whose production needs both natural and socio-economic capital. The natural resources are environmental components, such as soil fertility and the necessary water supply for irrigation, while the socioeconomic resources include labor and fertilizers. Researchers have been looking at new food sources, tweaking existing ones, and even creating entirely new foods. So, what could be on our dinner plate thirty years from now?

And in addition to moving diets away from resource-intensive foods, like meat and dairy, the action is needed in order to improve governance and government food policy globally, to improve the affordability of food and increase universal access to it, as well as reduce food waste throughout the entire food supply chain.

The general aim of food companies is to produce familiar dietary staples, ones that are both cheaper and healthier, and do it as sustainably as possible, a word that we will get to know very well later in the book. Besides, substitute foods like vegan meatballs, mushroom foie-gras, coconut yogurt, and wheat-free flour have proved to be very profitable. Several "new kinds of milk" have found their ways to the market, such as almond, coconut, soy, hazelnut, cashew, hemp, and oat, and have been so successful that the dairy industry has started to feel threatened and protest concerning the use of the word "milk."

The truth is that people are more willing to eat substitutes for what they already know and love than trying something entirely new. It is hard to get populations to eat something that they don't recognize. However, another truth is that taste is a learned behavior. Yes, people can be taught and used to like all kinds of new things, including new foods.

Overall, the mainstream course is that instead of concocting new foods, flavors, and textures, food-related innovations are overwhelmingly focused on experimenting around the reinvention of what we already know and like. The same goes for new food technology; food-tech companies are claiming to create the food of the future, yet they rely on ideas of the past and present. For instance, cashew cheese is not technically cheese, yet that's what we call it.

Food science and chemical engineering are two scientific fields that are making realities today – from food pills to new species of built-from-scratch fish. Food science startups are experimenting with inventing meatless burgers by growing beef hamburgers from cow cells in Petri dishes (those shallow transparent cylindrical dishes that biologists use in their labs to culture cells) or creating a synthetic kind of egg that they will be able to use to lift muffins, suspend mayonnaise, bind cookies and everything else that a real egg would allow them to do.

The changes, of course, don't end in our kitchens and the food companies. Agriculture must follow as well, dynamically, and revolutionarily. The planet's arable land is limited, and deforestation is more harmful than beneficial. At the same time, **fresh produce** is indubitably the best produce; nothing beats the freshness of the crops that you've just harvested concerning the nutritional value and other health benefits. So, with that in mind, the sector's answer to the future is the incorporation of vertical farms into our grocery stores. With this cultivating method, we'll be able to modernize agriculture as well as maintaining the benefits of fresh local produce with limited food miles.

Business-wise, the future of food companies seems to be found in increasingly **bigger and stronger mergers**. New technologies, coupled with new business models and supportive government policies, can help develop more flexible urban food-ecosystems. Such tech-enabled systems can increase overall food production as well as generate opportunities for new businesses and jobs. If those food ecosystems are interlinked thanks to new technologies, small farms will be able to reach wider markets and progress from subsistence farming to commercially producing niche cash crops and animal protein, such as poultry, fish, pork, and insects.

Photo by Alex Gruber on Unsplash

THE CURRENT CHALLENGES

WORLD HUNGER

Worldwide, nearly 800 million people suffer from malnutrition. Hunger is a major problem, particularly in developing countries, as the majority of people affected by hunger live in rural areas of such countries. The latest international statistics on malnutrition make for uncomfortable reading. The populations in industrialized nations, by contrast, are increasingly affected by obesity: in the period from 1980 to 2014, the number of obese adults in these countries more than doubled to over 600 million (WHO 2016, Welthungerhilfe 2016).

Demand is continuously growing; by 2050, we will need to produce 70% more food. The over 500 million smallholders around the world are responsible for half of the world's food supply; in developing countries in Asia and sub-Saharan Africa, they manage and are responsible for as much as 80% of the farmland. However, they are less productive than agricultural operations in industrialized countries (FAO, 2014). Roughly 30 percent of the global workforce is employed in agriculture. That is approximately one billion people (ILOSTAT 2016).

Marion Nestle, Professor of Nutrition, Food Studies, and Public Health at NYU, does not seem to be particularly optimistic when it comes to the future shape of how the world is fed. On the website Time.com, she stated: "Unless there are big changes within the next 20 years, I foresee a two-class food system. One class will eat industrialized food produced as cheaply as possible at the expense of its workers and natural resources. The other will enjoy home gardens and locally and sustainably produced food, at a greater cost. I'm hoping for the enormous expansion of this latter approach. For that, we need a farm policy inextricably linked to health and environmental policy. We can achieve that, but only with serious advocacy and political engagement."[1]

DEMOGRAPHIC & DIETARY CHANGES

In the coming decades, the global population is expected to grow by 33% and reach almost 10 billion by 2050, up from 7.6 billion in 2017. But when it comes to demography and food, the numbers do not tell the whole story. **Firstly**, such a high population growth will exponentially up demand food by roughly 50% as compared to the 2013 agricultural output.

Secondly, the global diet itself is changing as well, because **thirdly**, global population growth comes accompanied by a global urbanization wave. In fact, global urbanization between now (2019) and 2050 could lead to a net addition of 2.4 billion people in towns and cities. Urbanization brings changes to the world's trade routes, especially to those of perishable goods, such as food products. It grows the demand for high-value animal protein. Urbanization also raises income, which in turn increases the demand for processed and prepackaged foods.

Something that must not be understated under any circumstance is the health burdens brought about by the excessive consumption of meat. And in developed societies, several types of meat enjoy high demand. As incomes increase, many people shift to diets where meats, dairy, oils, and refined carbohydrates are predominant. Also, fast-food restaurants usually thrive, and the availability of fresh foods gets limited. All three mentioned types of food (meat, processed, and fast food) lead to many chronic health problems, such as diabetes, high blood pressure, and heart conditions, as well as to a crisis in childhood obesity.

In succession, increased meat production bears negative effects on the environment. Raising livestock equals to almost 25% of all global water use in agriculture and is responsible for an estimated 18% of greenhouse gas emissions

---

[1] The Future of Food: Experts Predict How Our Plates Will Change, http://time.com/3482452/future-of-food/

caused by humans. In the long term, the impact on the environment is certainly unsustainable.

If the demand for food increases in the future due to the global population's increase, then this will be reflected in higher food prices. The system of sustainable intensification that we'll discuss later is a mechanism that could properly respond to that new need. It incorporates farming techniques that allow food producers to respond to those price changes. Of course, different regions will need to balance their goals as they best meet the local needs.

CLIMATE CHANGE

Photo by Markus Spiske on Unsplash

Climate change is rapidly altering the environment. Extreme weather events are becoming increasingly common, including record-high temperatures, floods, and droughts. One side effect of climate change is a rise in such phenomena which negatively affect crop yields. And it has been documented that crop yields decline significantly when daytime temperatures exceed a certain level (FAO, 2016e).

Pollution, deforestation, landfill waste, clean water insufficiencies, and other environmental effects of human activities have all reached all-time highs in recent decades. In particular, the degree of human-made emissions of greenhouse gases (GHGs) has reached the highest in history, according to a 2014 report of the Intergovernmental Panel on Climate Change (IPCC). Many species of mammals, birds, fish, and plants are at risk of extinction.

Agriculture is one of the primary producers of GHGs. Specifically, over the past 50 years, GHGs resulting from agriculture, forestry, and other land use have nearly doubled. ***The sector of agriculture contributes the largest share of global methane and nitrous oxide emissions***. High-consuming countries like the US and the UK must prioritize shifting toward predominantly plant-based diets in order to meet the international Paris Agreement on climate change.

Internationally, there has been a rapid uptake of the term Climate-Smart Agriculture (CSA), as food production enterprises aim at developing a safe operating space for their crops. The food system and the climate are two hugely interconnected systems. So, since we can't do much about the latter, the crops of the future will be indoors ones, as we'll see later in the book.

It must be acknowledged that climate-smart interventions in any food production system require much-specialized knowledge and can only be applied in a defined location. One of the things that make agriculture such a vulnerable sector is that several nutrients (e.g., nitrogen, phosphorus, and potassium) are required in crop growth – and they're all required in different quantities. Plants need to be well-supplied with nutrients during the entire growing period to reach their potential production levels. So, the slightest change in climate immediately starts messing up this sector.

Moreover, two sectors that usually lack the financial resources needed to implement innovative solutions are forestry and biodiversity protection. **Rainforest Connection**, a San Francisco-based startup company, transforms recycled cellphones into solar-powered devices that are attached to trees and detect logging activities at a great distance, aiming at combatting illegal deforestation.

Also, we can save a whole lot of energy with the help of monitoring its usage. **Nest**, another startup, has developed the smart thermostat that learns a household's patterns and automatically adjusts to them to save energy. Similar IoT applications can "smartly" monitor our sprinkler systems to save water, turn off our lights to save energy, and suggest to use different routes when driving to save gas (and avoid traffic, as well).

Given the above, it's easy to assume that the topic of food and climate change are pretty unrelated. If you ask around, most people – including me, until recently – will tell you that the biggest threat of climate change will be the floods caused by rising sea levels. Or, depending on the area, it might be drought (that is, the exact opposite!), the necessary displacement of populations, etc. But no! Actually, the single biggest threat that climate change faces is the disruption of food systems. Plant-based foods have a lower environmental impact than meat and dairy. Subtly in some cases, tremendously in others, the disruption of the food systems will be the one severe issue that is going to affect everybody on the planet at the same time.

And it's not just an issue for the coming years. Climate change has become something that we can taste. Farmers around the world have experienced differences in the way that their cattle, fruits, and vegetables grow. And not only that. Climate change is bound to affect every part of the food production chain, including food supply, food access, and, surely, food quality.

However, as we saw in the examples a couple of paragraphs above, it's not all bad news. The good-news response has been given by the Internet of Things (abbreviated as IoT) and, more specifically, thanks to the currently developing arts of **detecting** and **monitoring**. Thanks to human ingenuity, we are finding new ways to adapt. That is also the source of hope and optimism that we need when faced with this issue.

Investments in information about agriculture, ecosystem services, markets, and human populations must be structured not only to improve single parts of the food system but also to identify limits and provide practical guiding lines toward a sustainable future. Keys to this movement are better water management and careful matching of crops to environments. (Both are further discussed in their respective chapters.) The global agricultural community needs to move toward the "safe operating space," which I mentioned earlier in order to provide adequate food for the entire population without crossing critical environmental thresholds.

From robotic weeders and farmscrapers to genetic modification and artificial intelligence, we have examples of innovation and adaptation that can easily combat any feeling of despair or desperation, which are not helpful either way. So, our noteworthy **adaptability** will allow us to keep on producing the foods that we've been eating and enjoying for many generations, but with a twist. And that twist will be the growing method. We'll probably be eating equally delicious meat, except it won't have come from a live animal but will have been cultured in a bioreactor or will have come from a plant-based protein.

LAND DETERIORATION

The amount of arable land available for food production per person is limited and constantly decreasing. This is due to population growth, but also due to factors such as urbanization, erosion, and desertification. According to GM (Global Mechanism), 25% of all farmland is already rated as "highly degraded," while another 44% is "moderately" or "slightly" degraded. More than 200 million hectares of soil in Latin America are severely damaged (WRI 2016).

So, the land has long been recognized as a finite resource. What's more, water resources are alarmingly strained; more than 40% of the world's rural population living in water-scarce areas. As expected, the land shortage has led to smaller farms and lower production per person, both adding to rural poverty. Shortened fallow periods had turned shifting cultivation into a non-sustainable method, even though it used to be the opposite when population densities were lower.

Photo by MILKOVÍ on Unsplash

Fertile soil is being lost all over the world, due to factors such as deforestation, overgrazing, and mismanagement. When the use of fertilizers is unbalanced, it leads to an imbalance in nutrients. The vast majority of deforestation (of unsuitable land) indeed takes place for agricultural reasons. Now, while clearing vegetation to make way for farmland does not directly cause degradation of the soil, it does so indirectly by eroding water resources. Also, water erosion has been caused by the overcutting of vegetation, which has equally brought about wind erosion, resulting in land less suitable for food crops.

Agroecosystems can actually produce negative outputs. For instance, poor soil management can lead to reduced fertility; nitrogen run-off from agricultural land pollutes lakes or the sea; carbon gets sequestered in the soil. Then, there's the issue of overgrazing that directly leads to decreases in the vegetation cover's quantity and quality. This in turn compromises the soil's physical properties and resistance to erosion.

For financial reasons, farmers have turned to cereal-based, intensive crop rotations instead of more balanced rotations. The decisions that farmers have to make affect not only their livelihoods but also those of many other people. Indirectly, they affect the wealth, health, and wellbeing of much wider groups of people.

FOOD PACKAGING

Photo by Arabica Kyoto Arashiyama, Kyōto-shi, Japan

It's not just food production and the utilized first materials that we must adapt as soon as possible. Food packaging needs to become environmentally friendly and sustainable, too, especially concerning the use of plastic. It's part of the food production chain, after all. In fact, the packaging is fundamental in helping the prevention and reduction of food waste, as it protects food products from damage, spoilage, and contamination. After millions of tons of plastic debris (such as bags, wrappers, and containers) thrown in the sea, this is long overdue!

As more and more shocking images have come out depicting oceans filled with plastic waste and suffocating marine wildlife, people's opinions have shifted concerning the use of plastic packages and plastics in general. In turn, as consumers get more and more environmentally conscious, companies – not just food companies, of course – have no other options any longer.

It is almost certain that the future of food packaging will be plastic-free. There will be plastic-emulating materials instead. Currently, the main focus is on biodegradable products, while there's also some focus on developing edible packages. Another solution is the refilling of certain products, such as cleaning and hygiene products.

TIPA is a startup whose aim is to create viable solutions when it comes to plastic packaging. It looks to create a compostable and recyclable package that, once discarded, will decompose and leave no toxic residue behind it. It reminds me of another new creation, the edible coffee cup, that you consume after finishing your coffee or tea, or whatever it is you drank! Also, **Skipping Rocks Lab**, a London-based startup company, has created an edible – yet tasteless – water bottle (!) made from seaweed pouches.

New opportunities for packaging in many areas and not just concerning food products will open thanks to printed electronics and sensor technology. We have already witnessed this with the use of QR codes, which are just the beginning, of course. They are the ideal way to enhance food packaging. They turn it from a simple wrapping product into something much more powerful.

When it comes to food, companies can offer useful information about the ingredients, the food's nutritional value, or the packaging via QR codes. Aside from food-related information, those codes can link you to anything that a company deems will increase their sales.

For instance, in Hong Kong, Pizza Hut turned its takeout box into a mobile movie projector. They bundled a projector lens with their pizza and a perforated window in one wall of the box and added a code for the customer. The customer scans the code, places their smartphone inside the box, and can stream a movie using the pizza box as a casing for the projection. In the US, Taco Bell made a more targeted move during the College Bowl Championship Series; they added on their packaging QR codes that linked to a sports analyst previewing the next big game.

So, we can easily see that the possibilities of the use of QR codes are extremely broad, limitless even. They can significantly help numerous types of businesses become more appealing, especially to tech-savvy consumers. And this is not exclusive to big businesses. QR technology offers a chance to small and medium businesses to instantly differentiate themselves right from the get-go.

FOOD LOSS

Food waste is a massive market inefficiency, a kind of loss that doesn't exist in other industries. In the book's introduction, we talked about the problems of hunger and undernourishment on a global level and how about two billion people in total suffer from one of the two. Well, all those people could be fed with less than a quarter of the UK, US, and Europe's yearly food waste![2]

Every year, more than one billion tons of food are lost worldwide. **Undernourishment doesn't happen because we don't produce enough food to go around, but because of enough food, we produce doesn't go around.** In a word, logistics. Food loss can be encountered along the entire way from the fields to our plates. In poorer regions, food is mainly lost during production and storage, while in richer regions, losses are encountered, because consumers throw away large quantities of food.

In industrialized nations, consumers are responsible for most of these losses: 13 percent of the food purchased in Europe ends up in the garbage, while in the United States, this figure is almost 16% (Food and Agriculture Organisation, 2011). A Food Sustainability Index research used data from the UN's FAO and found that richer countries perform more poorly in the FSI in terms of food waste. The numbers are scary: In the US, annual food waste per headstands at 95.1kg. in Belgium, that's 87.1kg, and in Canada, it's 78.2kg.

Food waste is *bad for human hunger* and *bad for the economy* but also *bad for the environment* on two different levels. Firstly, there's a whole list of resources that are wasted away since the food was uneaten. Those are land, water, labor, energy, manufacturing, and packaging. And secondly, since most of the food waste ends up

---

[2] Food Waste Facts, https://www.tristramstuart.co.uk/food-waste-facts

in landfills, it decomposes without access to oxygen, and as a consequence, it creates methane (which is 23 times more effective in carrying out the greenhouse effect than carbon dioxide!)

Now, when it comes to concerns about how we are going to feed humanity's increasing population in the future often focus on how to increase food production while limiting the use of natural resources and the processes' carbon emissions. But another way to solve this is by reducing the enormous amounts of food constantly wasted. So, **food loss can be tackled from both ends**.

How? The magic word here is "**policy**." High-income countries must develop national food waste strategies. The good news is that many of them, including France, Italy, Spain, and the US, have already done so. In France, new food-waste-related legislation requires supermarkets with a surface of at least 400 square meters to redistribute leftover food to charities.

Similarly, several other countries, such as Australia, Canada, Slovenia, Sweden, and the US, have engineered laws, regulations, and regulatory instruments to make retailers donate unsold food. But it's not done with force; the donating process is also facilitated, while the relevant laws reduce those retailers' liability. And why not **give away** something that you were going to **throw away** in the first place?

Let's not just stay on the governments' end, though. The food sector, too, can innovate and develop products, processes, and business models that help tackle food loss. And the food sector has already developed several different ideas, promising major achievements.

For instance, a startup can help smallholder farmers get better access to advanced storage facilities. The UK-based startup **InspiraFarms** has developed off-grid cold storage equipment for farmers in emerging markets.

In sub-Saharan Africa, a startup company called **Wakati** has developed a hydration tent powered by solar energy. With that, farmers don't need cooling technology. The tent eliminates the production of ethylene, which accelerates the fruit and vegetables' rotting process.

The **Massachusetts Institute of Technology** uses food sensors to reduce food waste. Those sensors let you know when any food is going bad, and they do that at a very early stage, so with great precision, you can consume first or use as an ingredient the materials that would normally go to waste if it wasn't for the nanotech sensors.

In Japan, there's an app called **Reduce Go** that lets registered users collect unused food from restaurants and other food outlets for a monthly fee of ¥1,980 (US$17.50).

In Kenya, the startup **Twiga Foods** is an initiative aiming at establishing an efficient food distribution. Twiga has launched a smartphone application that consumers can use to place their orders in advance of the market day. The company has set up many warehouses in strategic points across the country, and this way, the product can stay fresh and be delivered promptly.

In the UK, a company called **Rubies in the Rubble** uses ingredients, mainly fruit and vegetables, that would otherwise go to waste to make condiments. Food produce can go to waste just for having a "bad" shape or the "wrong" size or color that we're used to seeing in the market. This company saves millions of fruit and vegetables from becoming food loss by using them to make an award-winning range of mayonnaises, kinds of ketchup, and relishes.

HEALTH PROBLEMS

The way we have been eating in the last two decades (that is, since around the turn of the millennium) causes several health problems, the most critical of which will be discussed in this segment. It is self-evident that the root of the problem hides in the ingredients of the most consumed food products. In recent decades, the leading cause of death on the planet is none other than NCDs. Noncommunicable diseases that one would expect to be less perilous than the contagious ones represent 70% of all deaths, accounting for the death of more than 41 million people every year.

According to conclusive data of the World Health Organization from 2018 for the year 2016, "Four groups of diseases account for three-quarters of [global] deaths: cardiovascular disease (17.9 million), cancers (9 million), respiratory diseases (3.8 million), and diabetes (1.6 million)," and compared to data from the year 2000, the numbers for all types of disease have been on the rise and, what's more, for all income groups, as well.

The sneakiest "bad guy" is a sugar that exists in one form or another in the majority of prepackaged food products. Speaking of ingredients in food products, the second enemy we need to face is the excessive presence of salt in other such products and sometimes in the very same ones. And then, there are the "bad" fatty culinary substances. Tobacco consumption kills more people than the diseases caused by excessive sugar consumption, and alcohol consumption comes in at number three on the malefactors' list. But what's particular about sugar is that it's hidden in too many food products, even in many snacks that are considered 'salty' and not 'sweet.'

Yet, despite the prevalent unhealthiness on the shelves of the supermarkets around the world, we can actually be very optimistic, because a large percentage of people have already acquired the habit of reading the foods' etiquettes and caring significantly more about what they put in their body and how it affects their health.

Sugar is certainly not the only one to blame.

When it comes to **sugar**, it's not the ultimate culprit in itself. It's its excessive consumption that leads to obesity, which in turn increases the risk of diabetes. Sugar is indeed more addictive than cocaine, while it's also associated with obesity and dental caries. Of course, obesity or overweight is not caused solely by sugar consumption. A sedentary lifestyle, disorganized eating, and sleeping regimes are also factors that cause weight gain.

Before we move on, something that must be pointed out is that total abstinence from sugar is neither a realistic nor a healthy option. All kinds of diets need – and have – sugar in it, simply because sugar is in some form contained in all fruits, vegetables, nuts, and dairy products. To completely eliminate it from our diet would involve us being stuck with consuming only fats and proteins, something that is not conducive to long term health benefits. Sugar is not a toxin. In moderate amounts, it is part of a balanced diet.

According to the April 2019 report of the Task Force on Fiscal Policy for Health, "More than 10 million premature deaths each year – about 16 percent of all deaths in the world – could be prevented by reducing consumption of three products: tobacco, alcohol, and sugary beverages." Of those three products, tobacco constitutes the biggest health risk globally, "accounting for 8 million deaths a year." The equivalent grieving numbers for sugar and alcohol stand at six million and three million, respectively.

Now, when it comes to the future of meat and dairy in our food, health, and the environment, things get noticeably complex and nuanced. On the one hand, food production coming from animals majorly contributes to greenhouse gas emissions and are excessively consumed, yet on the other, those food products are rich in micronutrients and will soon have a key role to play in combatting global malnutrition.

OBESITY

According to statistics of the World Health Organization (WHO), obesity on a global scale has nearly tripled since 1975. One in three adults around the world (roughly 1.46 billion) were overweight or obese in 2008, that is, up by 23% since the year 1980, less than three decades ago. Then, in 2016, the number of overweight adults worldwide surpassed 1.9 billion in just eight years, with 650 million of them being obese.[3] And according to data from the CDC National Center for Health Statistics, 39.8% of adults aged 20 and over in the US have obesity for the years 2015-2016.[4] This is more than one-third of adult people in a country with a total population of over 300 million citizens!

In the developing world, more specifically, the number of overweight or obese adults more than tripled from 250 million in 1980 to 904 million in 2008. The United Nations' statements are pretty much the same. In late 2016, they stated that there are more obese than overweight people in the world, with children and adolescents' obesity rising rapidly, particularly in developing countries. Even more gravely, when it comes to associated health issues, obese children may experience breathing difficulties, increased risk of fractures, hypertension, early markers of cardiovascular disease, insulin resistance, and psychological effects.

Numbers-wise, in Africa, the number of children who are overweight or obese nearly doubled to a stupendous 10.6 million in 2014 from an already worrying 5.4 million in 1990. In Europe, Britain has the highest level of obesity in Western Europe, ahead of countries such as France, Germany, Spain, and Sweden, according to a 2013 FAO report. Obesity levels in Britain have more than trebled in the last 30 years, and, on current estimates, more than half the world population could be obese by 2050.

I know that reading a bunch of numbers might be a little tiring, but the sad fact is that a part of those numbers will possibly translate into deaths. This is large because most of the world's population lives in countries where problems of being overweight or obese kill more people than problems associated with being underweight. Obese or overweight people are at risk of several diet-related diseases, including cardiovascular disease, diabetes, strokes, even some types of cancer. Not

---

[3] World Health Organization, Obesity and Overweight, https://www.who.int/news-room/fact-sheets/detail/obesity-and-overweight
[4] CDC Centers for Disease Control and Prevention, Overweight & Obesity, Data & Statistics, Adult Obesity Facts, https://www.cdc.gov/obesity/data/adult.html

to exclude that some of them may also suffer from micronutrient deficiencies. Obesity is on the rise in urban areas in low- and middle-income countries. And 3.4 million people die each year due to their being overweight or obese.

The increasing consumption of sugar and sweeteners has risen by over 20% per person between 1961 and 2009.[5] As we drive more and more rather than walk or bike, as we are increasingly stuck in front of a screen and order online, as we exercise the combination of body, mind, and soul less and less, we get fatter and fatter, as one would easily assume. Looking at the food consumption statistics of governments around the world, we can see that in recent years, people have gotten used to eating much differently than in the past. We now buy less fresh milk, but more ice cream and an alarming amount of dairy desserts, all of which are filled to the brim with sugar.

We also buy much more cereals and cereal snacks, which are once again coated with sugar, mainly to make children adore them and become addicted to such products. Equally alarming is the fact that we buy fewer potatoes but more potato chips. Another interesting thing is that we buy less sugar in its standard form, yet we consume dangerously more of it. You see, instead of making and baking in our kitchen on our own, following our grandma's recipes or those in cookbooks, we prefer to buy snacks, drinks, sodas, sweets and all kinds of confectionery, all of which – that's right, you guessed it – contain an insane amount of sugar.

This unprecedented consumption of the various forms of sugar seems to be the main concern on a global scale when it comes to what has caused the current high levels of obesity in the majority of countries around the world. All these changes did not take place by accident; food companies have contributed to the problem bigly it by investing in lacing products packed with addictive sugar, in turn, harming many who lack the knowledge to make an informed judgment on their foods nutritional value as well as by employing advertisers, psychologists, and food scientists to help them develop and promote food products as well as a rhetoric that

---

[5] Food Consumption Trends and Drivers, https://www.ncbi.nlm.nih.gov/pmc/articles/PMC2935122/

will convince us to eat more than how much we truly need. And combined with an unjustifiable lack of nutritional education on the comfortably numb consumers' part, those companies had it comfortably easy.

On the contrary, the weak-of-heart had a tough time resisting, and unrightfully, the blame fell upon them. It is like intentionally brainwashing people only to blame than the ones who got brainwashed. A noteworthy immoral aspect of all this is that people who are not in the best psychological or mental state were taken advantage of. Food companies fed them with ingredients that acted like drugs and were fed off of those people's almost inescapable addiction. But I'll stop here because this is a book about the future and not a trial about the corporate past.

Lately, there has been another shift in the food trends, with healthy snacks filled with proteins, plant fiber, and vitamins instead of sugar taking the place of the unhealthy ones of earlier years. Yet, the health damage on the masses has already been done and it still needs to be addressed and fixed. "In most rich nations, obesity rates are much higher at the bottom of the socioeconomic scale. They correlate strongly with inequality [...]. The scientific literature shows how the lower spending power, stress, anxiety, and depression associated with low social status makes people more vulnerable to bad diets. Just as jobless people are blamed for structural unemployment, and indebted people are blamed for impossible housing costs, fat people are blamed for a societal problem."[6]

In Chile, a country of 18 million citizens and overloaded with obesity, the government chose to tackle the issue vigorously and head-on. So, what they did was impose an "iron-fist" legislation against the current unhealthy foods and establishing new eating habits for the people, especially at schools. The "fight" began in 2011, but the laws were adopted in 2016 with the President of Chile Michelle Bachelet, who is also a podiatrist. A culminating point was when it was prohibited for ice creams, chocolates, and potato chips to be sold at schools. Fifteen hundred food products had to change their composition in order to avoid the

---

[6] We're in a New Age of Obesity. How Did It Happen? You'd Be Surprised, by George Monbiot, https://www.theguardian.com/commentisfree/2018/aug/15/age-of-obesity-shaming-overweight-people

obligatory black etiquette on their packaging, etiquettes that draw attention concerning the excessive quantities of sugar and unsaturated fats.

If our goal is to combat obesity, the necessary first step is to study and grasp the so-called "science of hunger fully," that is, the way our brain neurons work when we eat and/or see and/or smell food. The potential for combatting obesity lies exactly in clinically manipulating those neurons; that is the knowledge that could lead to the formulation of actual treatments for this epidemic.

The exploration is and will remain focused on the biological mechanism – of not just humans, by the way – behind the feeling of hunger. More precisely, when our energy stores are running low, the levels of our "hunger hormone," whose official name is ghrelin, begin to rise. But that is not the only eating-related hormone. We also have another kind of hormones, the peptides, that tell our brain whether we need to eat or feel full. In the first case, we're talking about the hunger-stimulating peptides (officially, orexigenic) and, in the second, about the hunger-suppressing ones (officially, anorexigenic).

DIABETES

For obvious reasons, everything that has to do with medicine, health, or everything in between is not considered sexy. If one writes something about that, one has to start with a shocking statement in order to draw people's attention. So, for this chapter about diabetes, here's the obligatory dramatic statement: According to **2016** data of the World Health Organization (WHO), over 400 million people have diabetes, with the rates increasing worldwide each year. And according to **2018** data, again of the WHO, over 1.6 million people die as a result of diabetes every year.

Older data from two successive years can also demonstrate how alarming the situation is: According to **2012** data of the International Diabetes Federation (IDF), about 371 million people were living on this planet who suffered from diabetes, and about 187 million of them were undiagnosed. That came after **2011**'s estimate, which was 366 million. An additional five million people within one year is not something to overlook.

Also, approximately 30 million people have diabetes in the US. And what does this Federation expect for the future? Well, by the year **2030,** which is not far at all, they expect that about 552 million people will be suffering from this diet-related disease.

Hundreds of millions of people have diabetes, half of them undiagnosed.

In recent years, there has been more and more talk about the good fat that exists in certain foods. Fat may be associated with extra weight, but we all know that our body needs a certain quantity of it, so the distinction between those two types of fat is the explanation of this dichotomy. The 'good' or brown fat is near our necks and shoulders and helps in our body's calorie-burning process; the 'bad' or white fat is found in the areas you probably would expect: for example, hips, thighs, stomach, etc. and unfortunately, it does not do much more than just storing calories, burdening our bodies and, by extension, often our psychological status as well.

Given all that, scientists have turned the focus of their research on converting white fat into brown – like a clinical penitentiary turning the bad substances into good ones.[7] The findings of the Washington University School of Medicine study – that was published September 19, 2017, in the journal *Cell Reports* – make it seem more possible than ever to develop more effective treatments for obesity and diabetes-related to weight gain. "The researchers found that blocking the activity of a specific protein in white fat triggered the fat to begin to brown into beige fat, a type of fat in between white and brown. Blocking the protein to create beige fat caused the fat cells to heat up and burn calories."[8] As the paper's first author, Irfan

---

[7] Cell Reports, Mature Human White Adipocytes Cultured under Membranes Maintain Identity, Function, and Can Transdifferentiate into Brown-like Adipocytes, https://www.cell.com/cell-reports/pdf/S2211-1247(19)30342-0.pdf

[8] Scientists Find Way to Convert Bad Body Fat into Good Fat, https://www.aau.edu/research-scholarship/featured-research-topics/scientists-find-way-convert-bad-body-fat-good-fat

J. Lodhi, Ph.D., said: "Our research suggests that by targeting a protein in white fat, we can convert bad fat into a type of fat that fights obesity."[9]

A major factor for today's high rates of diabetes occurrence is the excessive consumption of sugary beverages. This type of beverage causes insulin resistance directly and weight gain indirectly. Their consumption must be lessened to, in turn, lessen the added sugars in our diet.

Photo by Monika Grabkowska on Unsplash

---

[9] Ibid.

THE FUTURE AROUND FOOD

VEGANISM

The use of animals as food production technology is one of the main factors that have brought us to the verge of environmental catastrophe. So, it is high time we did something drastic about it. Today, food production is quickly moving out of agriculture and more and more into the tech domain. Where there used to be an animal-derived product, in the future, we will only find research, innovation, startups, launches, marketing, and sales.

Non-dairy milk was the first made-for-purpose vegan product to make a splash in the market. Soymilk had beef meat around for years when almond milk arrived in major supermarkets, boosting interest in the sector around 2012. Health, animal welfare, and the environmental impact are cited as the top reasons for switching to non-dairy, which now accounts for 12% of global milk sales. Another company, PG Tips, has even redesigned its tea bags to function better in non-dairy milk.

Biotech startups with the technology to extract, hack, or create sustainable protein will be the food lords of the future. Much time and energy are being spent producing vegan foods that look and taste just like their meat-based originals: sausages that sizzle, steak with "chew," fish goujons that are firm but flaky. In fact, the target market for much vegan food and dairy is the so-called "flexitarians": those who aim for a plant-based diet but allow themselves the occasional meat dish.

Moreover, in a San Francisco lab, Memphis Meats, co-founded by cardiologist Uma Valeti, is taking animal stem cells and growing meatballs and chicken. Then, another startup, Perfect Day, is using fermentation in microflora to make an animal-free dairy protein that could be used to make ice cream and other foods "without the need of a single cow." Hens are pretty much redundant, too, now that egg whites are created in the labs of Clara Foods.

In 2018, 16% of new food products in the UK were vegan, the highest percentage in the world. All the major supermarkets invested massively in their own brands. Tesco partnered with chef Derek Sarno to launch Wicked Kitchen and doubled the range when demand for plant-based ready meals soared by 25%. The convenience food sector has been invigorated by vegan machines, drive-throughs, and proliferation of plant-based ready meal delivery companies, such as London-based Allplants.

The giants of technology, from Bill Gates to Jeff Bezos, Twitter co-founders Biz Stone and Evan Williams, and Google co-founder Sergey Brin, have all invested in vegan startups. Another noteworthy remark is that Impossible Burger 2.0, a plant-based burger that bleeds real beetroot juice blood, was hailed as the best technological innovation at the Consumer Electronics Show in January 2019.

Unfortunately, it's not all good news. Using plants in food production technology isn't the perfect solution, because delivering Earth-friendly vegan products has created a fresh set of problems. Around 10% of California's water goes into almond production, while demand for soy, a big ingredient of plant-based meat, is driving the destruction of Brazilian grasslands. The answer is to make food from scratch.

AN INDIVIDUALIZED DNA-BASED DIET

Did you know that medicine to many diseases, including obesity, diabetes, Parkinson's, and multiple types of cancer, can be found in our own body? Yes, the human body's microbiome, that is, the trillions of microorganisms that reside in our intestine holds the key to the cure and prevention of many diseases, even life-threatening ones.

For centuries, one of the questions looming over humanity was whether the course of our life is mostly defined by nature or by our upbringing and lifestyle. At first, it was more of a philosophical question, but as the science of genetics progressed, it became a scientific one increasingly. In the course of the twentieth century, the conflict expanded to incorporate human intelligence as well as mental disorders, from aggression to schizophrenia. And depending each time on the new findings of the research, the index turned more toward science or philosophy.

But as the century approached its turn, the dilemma-looking question lost its significance as the field of epigenetics came to light. Epigenetics studies the modifications that take place in our DNA and that regulate if, when, and where a gene will be expressed, without altering our genetic code. Considering that our environment mainly determines these factors, the epigenetic studies showed that what we define as nature and raising interact so much with each other that it's unrealistic to want to calculate how each of them influences our development accurately.

From a philosophical point of view, that's very liberating. It frees us from the genes' domination and allows us, through the reshaping of our environment, to largely take our life into our hands. That liberation proves to be a big one, thanks to our gut microbiome. Our microflora, that sea of microorganisms inside our body, is richer in our skin and intestine. That microbiome defines, to a high degree both our physical and mental health. So, its study could revolutionize many areas of medicine.

On March 1, 2018, a big study was published in the accredited scientific review "Nature" by no less than 70 researchers at the Weizmann Institute of Science (located in Rehovot, Israel) which showed that it is the environmental factors rather than the genetic ones that influence more the composition and function of our intestine's microbiome. So, the way we "feed" our microflora is more important compared to our given nature.

We now know that the composition of our microbiome – which defines whether it will act to our benefit or not – does not depend on our genes; it has to do with environmental factors, the most important of which appear to be our diet and exercise. According to the study's findings, the genes' contribution to the microbiome's composition and function range between only 2% and 5%. Other influencing factors are smoking, the medicines we take as well as our sleeping pattern.

More analytically, that study included a sample of a thousand people, the biggest sample used for a microbiome study ever. And it was a very "enticing" sample in genetic terms since you can find all kinds of populations in Israel, from Ashkenazi Jews and African populations to Arabs. That genetic variety heterogeneity is like a geneticist's dream. Two teams of the Weizmann Institute collaborated; one was Professor Eran Elinav's team from the Department of Immunology, and the other was Professor Eran Segal's team from the Department of Computer Science and Applied Mathematics. One thing that the researchers discovered through this study was that they could group the participants genetically based on their origins.

But at the same time, people who belonged to the same genetic groups had completely different gut microbiomes, which clearly means that one's genetic background is not associated with one's microflora. The researchers also studied people with genetic similarities, such as parents and children, and people who lived together without common genes, such as partners. They found many more gut similarities between people who shared the same diet and lifestyle than those who were just related and not living under the same roof, thus not eating the same meals.

Various metabolic factors contribute to the occurrence of diseases, such as diabetes, fatty liver, cardiovascular disease, and obesity. Modifying our microbiome probably means that those factors will also be modified, so the outcome of those diseases could then change (for the better). Our gut microbiome influences various inflammatory diseases, like the inflammatory bowel disease, as well as on autoimmune diseases that can 'hit' even organs and systems far away from the bacteria's 'residence.'

The list of diseases that appear to be linked to our almighty microflora grows longer and longer, as different research groups around the world bring to light new connections between the bacteria in our intestines and the state of our health that sometimes seem quite unexpected. Equally, some characteristics of ours, like the perimeter of our belly, our glucose levels, and our BMI, are strongly linked to the intestine's bacteria.

Moreover, to provide a full perspective of the matter, our microbiome can be seen as a vast ecosystem of viruses and bacteria, as a giant biochemical factory, since those microorganisms produce or modify tens of thousands of metabolites that in turn – at least most of them – can travel in our body and have various effects on it. So, based on our new scientific knowledge, we have opened the road to the development of many new categories of medicines for numerous diseases. And that "base" is none other than our gut microflora.

Soon, we'll start getting our microbiome tested.

The emerging tools of scientific branches, such as genomics, proteomics, and metabolomics, allow us to study in-depth the relationships between diet, genetics, and metabolism. A complete picture of the human genome can lead food producers to the creation of foods that better meet our modern-day needs. It will probably be more expensive, but it will also be specifically customized to each individual and thus, much more beneficial.

There's all this information today about the way our gut and brain interact, and people – not just scientists – want to know what they can do with that knowledge to better their health. Prebiotics, probiotic foods, and psycho-biotics have entered our lives under the umbrella of the so-called functional foods, aka **brain foods**.

In about two decades from now, we may have sensors embedded in our intestines charged with the duty of tracking the health of our gut microbiome, aka our gut flora… Which, by the way, is highly revealing for a large number of diseases, infections, and vitamin deficiencies, among others. So, it doesn't sound bad to me. At all. If tracking all those trillions of microbes residing in our gut becomes so much easier (AND cheaper) in the near future, maintaining our precious gut health will be a piece of cake.

Every gut needs its prebiotics.

In recent years, there has been more and more talk about prebiotics. After all, they go hand in hand with our gut microbiome. Prebiotics are the food of the "good bacteria" in our intestine. Experts estimate that by 2050, supermarket shelves will have been filled with **functional foods**. These include yogurt, garlic, leeks, berries, onions, artichokes, and asparagus.

The good news is that we have discovered all that and can consider it while shaping the healthiest lifestyle that a person can have. The bad news is that we're probably going to need prebiotics (that is, supplements of them), especially those who take antibiotics, have a weak immune system, or are about to go into hospital or on holiday, independently of our diet. Otherwise, you see, we'd have to wolf down two onions or seven cloves of garlic every day. And that, to get the minimum amount of prebiotics our body needs. To see any improvement, we'd have to cram more than double those quantities.

But there is another piece of good news around prebiotics. Apart from helping us combat strictly health-related issues, such as obesity, insulin resistance, and cholesterol, the intake of prebiotics may even help us to feel happier, have a night of better sleep and reduce our levels of stress and anxiety.

Instead of just a baby food section, our supermarkets will have products tailored to every segment of the population – for instance, foods optimized for women, men, and the elderly. Food science will formulate the best nutritional profile for each demographic group, as well as for each individual.

DNANutricoach is a startup company dealing with the question "How can our DNA define what we eat?" and all its aspects. With a simple and fast method, the company can analyze each individual's genetic fingerprint and then design a personal guide with dietary directives and recommendations. This can give answers to basic issues, such as whether diet affects our genes, whether certain foods are beneficial for certain people and how we can protect our children who are genetically predisposed to a disease.

The company in question performs genetic analyses based on nutrigenetics. That is, it analyzes and estimates those genes which are exclusively associated with diet and exercise and which both affect and are affected by our eating habits. Part of the company's innovation is that it provides all its clients with advisory services that will help them achieve their health-related goals, which may not necessarily be diet-related. For instance, a goal may be to quit smoking. And part of its aim is to put each individual's diet in harmony with their genetic predispositions. The latter is achieved through food coaching sessions, either face to face or online.

Photo by <u>Dose Juice</u> on <u>Unsplash</u>

ECONOMY: THE SUGAR TAX

The logic of the 'sugar tax' is simple: If it's unhealthy, tax it! And if it's tremendously unhealthy, tax the hell out of it! You can help solve the diet-related health issues through information, through improving the nutritional content of foods, through a less 'chemical' agriculture, but at the same time, you can think in strictly economic terms, in market terms and adopt a complementary approach to this current global issue.

More explicitly, you can seek to contain or reverse this trend toward diets rich in sugars, oils, meat, and carbohydrates by regulating food supply or price, that is, by making healthy foods more affordable than the unhealthy food products. Governments (and multinational regulatory institutions) can also incentivize the production of a diverse range of grains, tubers, fruits, and vegetables.

You might not have thought so until now, but foods and food products are very hard to tax – properly, that is. There are quite a few factors that contribute to the estimation of a tax's height. And with food, things get incredibly complex. Food, that basic human need, marks perhaps the biggest inconsistencies, as it depends on where you are, what exactly you're eating, and how exactly you're eating it! Take cooking oils, for instance. In my native Greece, the VAT rate has been raised significantly, reaching 24%, which includes all cooking oils except for olive oil, a Greek cuisine staple. Olive oil's tax rate remains at 13%. But in Malawi, the 16.5% tax on edible oils was eliminated in July 2017.

As for certain food products, what matters is what exactly they contain. For example, in the UK, cookies, and cakes are exempt from VAT. But if you add a layer of chocolate, they immediately get subject to taxation. Another example comes from the US states of Illinois and Washington, where chocolate bars are exempt if they contain flour.[10] And finally, in many cases, what matters for the tax rate is whether you will eat what you ordered at the restaurant or buy it as a takeaway.

---

[10] Illinois Department of Revenue; "Some candy bars aren't candy," Seattle Times, May 31, 2010.

As a consequence of such complicated, messy calculations and exemptions, the idea of taxing luxuries and "guilty pleasures" is back, very much alive, and better than ever. After taxing alcohol, tobacco, and gambling, it's now time to tax food items like sugar and junk food. The main goal of the new excise taxes is to influence consumer behavior. For instance, sugar taxes on soft drinks are used to fight obesity and carbon taxes to address pollution.

The excise-taxed list keeps getting longer. It includes mobile phones, plastic bags, and foods, such as chocolate, coffee, and soft drinks. Now, the introduction of excise taxes is not limited to luxuries. With the help of easily collected data in the digital era, it's possible to quantify the negative effects of a society's usual behaviors when it comes to its consuming habits and shapes the excise taxation landscape accordingly in due time.

What we need to acknowledge is that certain behaviors (such as smoking or polluting) have a cost on not only ourselves but also other people, and that cost needs to be paid – and in a fairway. The excise taxes offer that way. And since the imposition of excise taxes helped reduce the habit of smoking, policymakers are now experimenting to find out what other bad habits they could reduce the same way.

Such measures are introduced by governments not to make money but rather to save money from the exorbitantly high cost public health-sector budget. Excise taxes get more and more popular with governments today around the world because they are easy to collect and relatively dissociated from recessions. Excise taxes are less affected by the fluctuations of the economy's cycles. Since a majority of people don't seem eager to adopt a healthy diet and lifestyle soon enough, eventually, the ball enters the governments' courts. And what else would they do but find a way to (financially) impose a healthier lifestyle to address their respective national expenses due to health issues of their population, such as diabetes and obesity?

On a side note, while those various tax rates will encourage consumers to turn to healthier foods or healthier versions of food products, they will also dynamize innovation in the food business and promote the market of food-production-related new technologies.

On the last note, excise taxes do have a potentially negative aspect, that is, partially harm a country's economy. This can happen if consumers of one country travel to a neighboring one that has not adopted such a high excise taxes or no excise taxes at all and buy their guilty pleasures from there. And while they're at it, they can do even more shopping in that country and thus, make their own country lose significant excise-tax revenue. Of course, once the new taxes become global, this potential negative effect will likely vanish into thin air.

A list of countries that have already established taxation on SSBs (sugar-sweetened beverages) is the following:

- Barbados (2015)
- Belgium (2016)
- Brunei (2017)
- Chile (2014)
- Dominica (2015)
- Finland (since all the way back to 2011)
- France (since all the way back to 2012)
- Ireland (2018)
- Mexico (2014)
- Portugal (2017)
- South Africa (2018)
- Spain (2017)
- UK (2018)
- US (2015)

Now, I realize that one could ask: "Okay, this sugar taxing sounds clever and all, but is it effective? Does it really work? Does it bring the desired results?" Well, let's have a look at the statistical data at our disposal. But before we do that, I'd like to point out a personal note. I find it unfortunate that all those taxes are called sugar tax, tobacco tax, alcohol tax, etc. when in reality, they are nothing but health-improving taxes. They have such a large positive health impact that they should be considered health taxes. They are designed, implemented, and imposed to reduce the frequency of nations' unhealthy habits and consequently, improve those nations' health status and, thus, reduce the overbearing healthcare cost that their states have to pay every year.

There is extensive evidence showing that the demand for all three unhealthy habits in question (tobacco, alcohol, and sugary beverages) is reduced when their price is increased. More analytically, as reported in the April 2019 report of the Task Force on Fiscal Policy for Health, "On average, in low- and middle-income countries, a 10 percent increase in price results in a 5 percent decline in tobacco consumption (NCI 2016), a 6 percent decline in alcohol consumption (Sornpaisarn et al. 2013), and a 12 percent decline in sugary beverage consumption (Powell et al. 2013)." So, the numbers are encouraging toward even higher excise 'health taxes.'

The decline of unhealthy consumption is even bigger among lower socio-economic groups and young people. For the former, higher prices reduce the long-term to expensive products, while for the latter, those prices can even prevent the youth's initiation of such consumption. Seeing the whole issue even more long-termly, the results look even brighter for the future, because children are in general likely to follow their parents' bad habits, so combatting those habits in the present will have a positive ripple effect for generations to come.

The effects that excise taxes bring on tobacco consumption can be seen even more clearly through specific examples of countries. In Brazil, between 2006 and 2011, the prices of cigarettes were increased by 34%, leading to a 19% drop in consumption per adult. Soon afterward, between 2012 and 2016, there was another price increase by 33%, bringing the consumption decline to almost 50%! (Iglesias 2016). In Russia, 2005 was the year when the Russian Federation first adopted several expanded alcohol control measures, such as banning advertisements, along with raising the alcohol tax rate. In less than a decade, alcohol consumption had been reduced by about 33%, bringing forth a significant decline in noncommunicable diseases and mortality rates.

Of course, aside from the positive impact on a nation's health, excise taxes generate a substantial source of income for the respective states. For instance, following the reforms on the taxation of both tobacco and alcohol, Colombia saw in 2017 an increase in tax revenues by a whopping 54% (for tobacco) and 17% (for alcohol), compared to one year earlier, in 2016. Also, after introducing a 'peso per liter' tax on sugary beverages in 2014, Mexico amassed 16 billion Mexican dollars (that's about 1 billion American dollars) in 2015.

Photo by Monika Grabkowska on Unsplash

FOOD PRODUCTION TRENDS

The current system needs to be disrupted, and the new technologies will allow for that to happen. Pathogens, pests, and weeds are increasingly threatening farmers' yields worldwide. Most initiatives aim at shifting the state of our food system from one that pigeonholes farmers into monoculture and floods our markets with refined soy, corn, and wheat to one that embraces sustainable, regenerative agriculture and policies for the improvement of public health instead.

Future agriculture will use sophisticated technologies, such as robots, temperature and moisture sensors, aerial images, and GPS. All those advanced devices and tools that will be discussed in this chapter aim at allowing farms to be more profitable, efficient, safe, and, of course, sustainable (which is largely synonymous to environmentally friendly).

We live in a fast-changing world, and farming is changing along with it. The next era of the agricultural sector also referred to as Agriculture 4.0, will see farmers using minimum quantities of water, fertilizers, and pesticides or even completely removing them from the supply chain. Instead, they will mostly use abundant and clean resources, like the sun and seawater, for their crops. Innovation can make an environmentally friendly, positive contribution to agricultural productivity. However, it requires a major investment in research and development.

New technologies are steadily taking over the food industry, bettering its value chain as a whole. According to Agfunder, agricultural technology startups have been growing more than 80% annually since 2012. On the walls of the hypermodern museum Museu do Amanhã (Museum of Tomorrow) in Rio, it says that global food production will need 50% more energy as well as 40% more water by 2050 to feed the planet.

A very auspicious food-production trend is placing sensors in trays of produce to continuously monitor the ambient temperature and sample the gases that fruits and vegetables give off, intending to calculate their remaining shelf life. This way, people who work in the food supply chain can know whether their product needs to be sold locally or if it can make a week-long journey and still be fresh enough to be safely consumed.

A French company, **Yoli LLC.**, has figured out how to use windmills to suck moisture out of the air. They're able to get somewhere around 500 liters a day, again depending on the humidity. A similar way to achieve this is by using solar power. The principal goal of this industry was to see who could create the self-filling water bottle, a bottle that will just sit there and fill itself. They have developed one that works on a bicycle! As you're riding along, it sucks the moisture out of the air and fills up the water bottle.

Another initiative that must absolutely be taken is to promote, market, and distribute the "ugly" produce. This will significantly reduce global food waste, as about a third of the world's food goes to waste due to its undesirable appearance – specifically, because of imperfections in shape or size. The "ugly" food waste is enough to feed two billion people, and we're talking about perfectly healthy and nutritious and equally tasty fruits and vegetables.

As more and more new inventions revolving around food production come up, as the demand for food diversity is about to be exponentially increased, as new opportunities are about to arise, and as the food market is going to shift dramatically, we will study in this chapter most of the existing newest inventions as well as those under development. Now, what do they all have in common? Many things: Sustainability, efficiency, personalization, fewer requirements in terms of time and effort, and obviously the best attainable quality possible.

THE CASE OF SUSTAINABLE INTENSIFICATION

"Sustainable Intensification" has been defined by London's The Royal Society as a form of production wherein "yields are increased without adverse environmental impact and without the cultivation of more land." Sustainable intensification is not linked to a single agricultural approach. It is based upon the principle that in a complex world with a growing population, sustainability requires the most effective use of inputs and reducing the undesirable outputs, intending to achieve greater yields.

It denotes a system of production that takes into account every resource and all information, including technology, natural capital, and land. It is land-intensive, which means that it makes intensive use of the given land rather than using a lot of land. It denotes a system of production that is environmentally sustainable as well as socially socioeconomically sustainable, a combination that also renders it an ethical system to promote and preserve.

The leading aim of sustainable intensification is to raise productivity while reducing any environmental impact. "This means increasing yields per unit of inputs (including nutrients, water, energy, capital, and land) as well as per unit of 'undesirable' outputs (such as greenhouse gas emissions or water pollution)."

The productivity increases to meet demand increases will be defined by the improvements in governance, waste reduction, dietary changes, and addressing population growth.

Aside from productivity increases, production increases will probably be needed in regions where agriculture is a crucial and underperforming force in their rural economies, such as in sub-Saharan Africa. Part of sustainability's aim is to set local goals of intensification instead of global ones. Not every place of the planet, not every country, not every kind of land can produce on the same level or dispose of the same resources.

So, today's main issues of concern are lowering agriculture's negative impact on the environment, the welfare of animals, and the quality and quantity of the agricultural sector's produce – all of which are issues that largely interact with each other. A sustainable food production strategy must aim at excluding non-renewable resources. There are only finite reserves of fossil fuels. Equally, the irrigation water stems from reserves like underground aquifers that are refilled so slowly that we could easily characterize them non-renewable too.

The concept of intensification includes the increasing use of indoor systems where food producers can have better control over waste emissions, food, water, and temperature. And the concept of sustainability includes the system's ability to withstand perturbation, whether humans or the environment causes that. Such perturbations include pests, diseases, the weather, and the state of the economy.

LAND SPARING

Land sparing means setting aside land for conserving biodiversity. This requires that the governance of land use is sufficiently sophisticated and operates on a sufficiently large scale that the bargain is met. The agricultural production on the cultivated land needs to be efficient to the maximum, in order for other areas to be left unexploited (for environmental and sustainability reasons).

Arable land and pastureland used to rear livestock could then be "spared" not only for environmental and other purposes but also to grow a more diverse range of nutrient-rich plant-based and tree-based foods. Land sparing belongs to the CA type (conservation agriculture), which was defined by the UN's FAO as "a concept for resource-saving agricultural crop production that strives to achieve acceptable profits together with high and sustained production levels while concurrently conserving the environment" in its 2007 report.

If we're going, being honest, it is difficult to really spare land for biodiversity in the face of possibly strong economic, social, and political interests pressuring on for its exploitation. For instance, political support, tax breaks, and subsidies provide considerable incentives for converting old-growth forestland. No matter the scientific evidence, without a trustworthy regulatory framework in place, the complex trade-offs underlying land sparing could probably never be made to work.

LIVESTOCK WELFARE

Photo by Leon Ephraïm on Unsplash

Animal welfare is becoming a more complex concept than what one might assume. It involves several different elements, but in general, animal welfare definitions incorporate two main requirements. Firstly, that animals are in good health, and secondly, that they get to experience a 'life worth living.' The 'good health' aspect of animal welfare is generally uncontested.

Then, disagreements concerning animal welfare often ensue, due to different people emphasizing those different elements to different degrees. In the developed countries, especially, the attribute of animal welfare is central when it comes to livestock breeding, as there must be compliance with the new animal-welfare legislation, which obviously comes in bearing higher standards. In developing countries, on the other hand, the demand for livestock products is on the rise, and animal welfare is a lower priority, so the respective laws are more lenient.

For all the differences that exist among countries concerning animal welfare legislation, it does constitute an issue that has gotten global. And the main reason why is the current globalization and international trade. What needs to be done is not simple; no matter how simple it does sound! So, what livestock-breeding businesses need to do is identify situations where both animal welfare and profit can be increased. Surely, to quantify and assess such cases requires an integrated assessing framework that allows the complicated measurements between animal welfare, efficiency, and financial outcomes.

TECHNOLOGY

Talking about the food trends of the future, the use of cutting-edge technology is certainly the main one. Some trends are possible, some are probable, and this one is certain — **the common denominator in all current evolutions and revolutions in technology. And the pace of change is only going to accelerate.** "The grocer of the future will rely less on stores, owning digital discovery and meeting customers on mobile to ensure revenue growth while foot traffic falls,"[11] according to Mark Collin, Group Director of Ventures at ThoughtWorks, a software consultancy enterprise. New technologies are here to solve current problems. If you're an entrepreneur, aspiring or existing, the main question to ask yourself is, "Am I truly fixing a problem, or am I really creating a new one?"

The global food system, and most notably the field of agriculture, is on the way to be reinvented with the aid of low-cost, high-tech methods for connecting food, people, tools, and useful data in vast global networks. Connectivity from simple cellphone SMS communication to Internet-enabled smartphones and cloud services provide platforms for increasingly powerful technologies. This is how the needed new agricultural revolution will take place.

Aside from connecting food producers to consumers in real-time, technology allows greater transparency and traceability throughout the value chains of goods and services in the market. And it doubtlessly allows a notable increase in profitability. Furthermore, blockchain technology allows greater access to basic financial services and the enhancement of food safety.

---

[11] ThoughtWorks | The Future of Food: The Stomach Wars Are on, by Charles Orton-Jones
https://www.thoughtworks.com/insights/blog/future-food-stomach-wars

Furthermore, food scanners are here to fix the issue that of our not knowing our food sufficiently. We can learn how many grams of sugar are contained in a piece of fruit with the help of **TellSpec** and **SciO**. The Nima sensor informs us about whether there is gluten in our food, while the Penguin sensor measures the level of harmful toxins.

Nutrigenomics helps us find out which foods we should consume and which ones we should avoid, strictly based on our DNA. It is an emerging science that studies how food affects our genes and how genetic variations affect our reaction to foods. Then, calorie counters help us decide when to eat or what to cook. Food chatbots aim at persuading people toward a healthier diet by telling us the most motivational lines imaginable! Such trends will be propelled forward with the help of technological advances, such as AI and real-time biomarker measurement (done by ingestible sensors or DNA profiles from saliva samples).

All in all, there are three main general trends around how technology is and will keep changing the food industry:

1. New food production techniques,
2. Cross-industry technologies and applications,
3. New technologies to bring food production to consumers, making the food supply chain more efficient.

3D PRINTING

Additive manufacturing aka 3D printing is a technology that has been applied in many sectors, such as architecture and biology, and can also convert alternative ingredients, like proteins from algae, beet leaves or insects into tasty and nutritious food products that can be produced in home kitchens by small, inexpensive printers. 3D-printed food can be customized according to not only a preference but also individual health needs, such as deficiencies.

It is a technology that turns a kind of gruel into a kind of meal. From the printer, it goes right onto the plate. We will be able to fully modify the shape, texture, taste, and form of our foods to match our personal expectations. For example, we will be ordering online our favorite chocolate bar or any other snack and then 3D-print it by using a piece of machinery that we will have at home. It sounds very appealing that we will be just a print away from the food we'll be craving at any given moment.

Experts believe that printers using hydrocolloids (substances that form gels with water) could be used to replace the base ingredients of foods with renewables like algae, duckweed, and grass. In the Netherlands, NOASR (which stands for Netherlands Organization for Applied Scientific Research) has developed a printing method for microalgae, a natural source of protein, carbohydrates, pigments, and antioxidants. They take those ingredients and turn them into something edible: into carrots, for instance. In one study, researchers added milled mealworm to a shortbread cookie recipe.[12]

HACKING PHOTOSYNTHESIS

Based at the University of Illinois, the RIPE project aims at reengineering the process of photosynthesis, which is the natural process found at the very heart of agriculture. What this regulation can achieve is highly increase crop yields without increasing the use of chemicals or the surface of needed land.

When it comes to photosynthesis, an essential enzyme is RuBisCO, found in plants' chloroplasts, where the process takes place. RuBisCO is responsible for the first part of the process of turning carbon from CO^2 into sugars. This enzyme can accept both carbon dioxide and oxygen; when it picks up the latter, toxic byproducts form.

That led plants to develop another process, which we named photorespiration. This second process dissolves the formed toxic chemicals and releases back to the photosynthetic pathway the carbon that they had clinched. Two problems there:

---

[12] Agriculture 4.0 – The Future Of Farming Technology, https://www.oliverwyman.com/our-expertise/insights/2018/feb/agriculture-4-0--the-future-of-farming-technology.html

Firstly, about 75% of the carbon is released, and secondly, photorespiration is a process that requires a lot of energy.

So, in comes genetic engineering. Scientists have already developed two new kinds of "shortcuts" to help the perfect realization of photorespiration. They applied one shortcut to the plant Arabidopsis (this is a common model plant) and one to the tobacco plant. Through alternative pathways, scientists aim at shortening the photorespiratory process.

Photo by Monika Grabkowska on Unsplash

SHARING SERVICES

Uberized services can promote development throughout the chain of urban food ecosystems, including production and distribution. Storage spaces and farming machinery constitute expensive equipment that can be shared, just like rides are shared through Uber and homes through Airbnb.

That includes the uberization of planting and harvesting equipment, transportation vehicles, refrigeration facilities for the temporary storage of perishable products as well as "cloud kitchens" that produce fresh meals to be delivered to urban customers. That way, young people who own motorbikes and cellphones have the chance to become entrepreneurs or contractors delivering meals to urban customers.

UNDERGROUND FARMING

There are already **underground hydroponic farms** that make use of LED lights, usually low-energy LEDs, instead of sunlight to grow plants. Hydroponics is a subset of hydroculture. It's a method of growing plants not using soil but mineral nutrient solutions in a water solvent instead. The plants are placed within an entirely controlled environment, where there is no wind or frost. They are grown throughout the year, unaffected by the weather and seasonal changes in an environment without pesticides. Also, certain crops can have LEDs of different colors if it helps their growth.

In south London, **Growing Underground**, an underground hydroponic farm was built in a network of abandoned tunnels, 33 meters below the surface. They have focused on leafy vegetables, from micro-herbs to baby leaf salad. The fact that it's not affected by the weather conditions is a huge plus for the company, considering Britain's unpredictable weather! Founded by Richard Ballard and Steven Dring in 2014, it now supplies local restaurants and retailers with fresh vegetables and herbs; it has successfully reduced the food miles and waste for local retailers and consumers, reconnecting the latter with local products. This initiative should become an inspiration across the board for abandoned warehouses, car parks, tunnels, and such spaces to be properly and effectively exploited and commercialized.

Also, cultures have started taking place within plantations. In northern Brazil, in the Amazonian state of Para, cacao trees and peppers are cultivated under the abundant shadow that only coconut palms can offer. Such techniques limit the soil's erosion and increase its fertility.

VERTICAL FARMING

Thousands of years ago, when humans started cultivating crops, they could only produce a given amount of food from a given area of land. It is with technology that crop density can become higher. So, one of the most important – if not THE most important of all – reasons why vertical farming's advancement is crucial is that if we don't increase production on the existing agricultural lands, then we'll have to convert more land to agriculture. And that is neither sustainable nor environmentally friendly, as one can easily deduce. Most of the land potentially available for conversion is forest, wetland, or pasture. Plus, the release of greenhouse gases is one of the land conversion consequences. Increasing yields is better for climate change than land conversion. So, the new needed policies should be based on the assumption that no land will be converted.

And it's not just the emissions of greenhouse gas. Conversion also endangers our planet's biodiversity. More specifically, the large blocks of forest in the Amazon and central Africa seem to regulate the rainfall patterns in a strikingly extended region around them in ways that should not be disrupted. Otherwise, if it happens, it would be very harmful to those two continents at first and the rest of the planet right after.

All that is problematic, so vertical farming moves in the exact opposite direction. It focuses on growing crops where suitable land is not available. It is the process of growing food in vertically stacked layers. It uses soil, hydroponic, or aeroponic growing methods. Vertical farming increases productivity while using an impressive 95% less water, fertilizers, and nutritional supplements, as well as no pesticides whatsoever.

Several new alternative production systems use "controlled environmental agriculture." These include poly hoop greenhouses, roof-top gardens, sack/container gardens, and vertical farming with artificial lighting.

Some of the vertical farming's most crucial benefits are that it allows to produce selected crops all year long, regardless of the weather conditions (which is increasingly determining due to climate change) and, since it avoids the need for loose farming soil, it does not deteriorate the soil conditions that could affect the crop's quality and productivity. In turn, it does not reduce the air quality because the unfarmed loose soil does not get blown into the air. Furthermore, it prevents agricultural runoff, which is one of the main causes of pollution in waterways.

The pair of underground and vertical farming gives us the main duo of the latest trends in the food field. Even though global yields have increased steadily in the last 60 years, this does not mean that we have any chance to meet future food demand without further expansion of the croplands. So, once again, we see that underground and vertical farming will actually become a necessity in the coming decades.

The advances being made in agriculture technology and the decreased costs of production can render vertical farming sustainable in the not too distant future. Currently, the valuable land for agriculture stands at 36% of usable land. With vertical farming, an agricultural enterprise can start using over 50% less land than what it used before. For instance, the rooftops of animal barns can be used to collect rainwater for grain and irrigation.

This is of utmost importance since, over the past half-century alone, farmland has grown by more than 400 million hectares, according to FAOSTAT[13] data. This conversion of natural habitats to farmland has also been one of the preeminent causes of seismic declines in terrestrial wildlife populations, which adds to the

---

[13] Food and Agriculture Organization of the United Nations.

importance and urgency to develop underground and vertical farming expansions. In 1900, 38% of Americans worked on farms; today, that number has fallen to just 2%! And let's not get sidetracked by focusing on farming. Megacities keep expanding, so these new ways of farming are but a humble step towards global conservation.

On the downside, vertical farming is dependent on affordable electricity. Because instead of sunlight, it requires the installation of special LED lamps to provide each plant with the necessary amount of light. But if we consider all its advantages, then governments should take the initiative and counterbalance that drawback. To make it cost-effective, governments can offer power subsidies or other tax incentives.

Under the direction of Japanese billionaire, SoftBank Vision Fund financed the San Francisco-based vertical indoor farming startup **Plenty** with USD 200 million. Other billionaires that have funded Plenty are Jeff Bezos and Eric Schmidt. In its field-scale indoor farms, 'Plenty' grows healthy food while minimizing the usage of both water and energy.

Another American company that has been building and operating indoor vertical farms since 2004 is **AeroFarms**. Its production is not affected by extreme weather events or seasonal changes. And its products are locally grown, not imported, so fruits and vegetables stay fresher for longer.

Governments have also initiated initiatives promoting the technology of vertical farming that renders the food chain more efficient. The Netherlands is one of the countries that have witnessed a boom in the indoor growing technique.

DESERT AGRICULTURE

As staggering as it sounds, one-third of the Earth's landmass consists of desert. So, aside from the seas that cover about 71% of the planet's surface, we need also to use the desert surfaces as food production places. It is an initiative that requires the combined brainpower of many different food scientists, specialized in different parts of food-tech and food production.

The frontrunner in desert agriculture is Saudi Arabia's KAUST (the King Abdullah University for Science and Technology). KAUST set off the Desert Agriculture Initiative, whose goal is to address and respond to the whole spectrum of tough challenges posed by farming in desert environments. Among other things, KAUST is working on genome-engineering technologies to manipulate biosystems aiming at plant growth, plant hormones, and growth regulators.

Approximately 60% of total desert agriculture productivity gets lost due to drought, salt, and heat; any improvement strictly depends on increasing the crops' abiotic stress tolerance. The ability of plants to adapt to extremer conditions hinges on the association with specific microbes and KAUST is determined to do just that.

Photo by Isaac N.C. on Unsplash

PRECISION AGRICULTURE & NANOTECHNOLOGY

Precision agriculture is another technological trend in the field of food that will see noticeable growth in the coming years. A new wave of innovators has brought it to the forefront. Precision agriculture is a set of technologies used to optimize inputs in order to maximize yields while having lower environmental impacts. It is mostly the developed countries that have seen gains from the emergence of precision agriculture.

The farm of the future is expected to be fully networked and dispose of a multitude of information assets. Experts will intelligently crosslink data from different sources with their IT centers, develop algorithms, and yield forecasts for each farm. To give one example, that kind of "smart" irrigation will significantly increase yields: Cutting-edge irrigation management can increase global kilocalorie production by 41%.

The Green Revolution of the twentieth century saw an unsophisticated application of pesticides and chemical fertilizers, whose consequences were a loss of soil biodiversity and a rise in resistance against pathogens and pests. Approximately 60% of the applied fertilizers are not applied, since they are lost to the environment, thus polluting it.

A new, sustainable, environmentally friendly revolution is needed, and that will be precision agriculture, driven by nanotechnology. More specifically, nanoparticles will be delivered to plants. Advanced biosensors will be used for precision farming. Also, Nano-encapsulated conventional fertilizers, pesticides, and herbicides will be used to release nutrients and agrochemicals in a slow and sustained manner, so that the plants get precisely the dosages they need each time.

Nanotechnology is used in the manufacturing of plant-derived coating materials that can decrease waste, extend shelf life and transportability of fruits and vegetables, and reduce post-harvest crop loss in developing countries that lack adequate refrigeration. Nanotechnology is also used in polymers to coat seeds to increase their shelf life and increase their germination success and production for high-value crops.

SATELLITES AND MOBILE RADIO ANTENNAS

In farming and ranching, people, crops, animals, and equipment are all spread out across long distances. Communication is key to coordinate all those elements, stay informed, and able to respond to changing conditions or even emergencies. Satellites will be used for more than just providing weather data. They will also help farmers measure the biomass of every different section of a field and then extract recommendations for action.

Satellite images will delineate any given field in detail and, combined with a GIS (geographical information system), they will allow more intensive and efficient practices of cultivation. This way, farmers' life will be facilitated as they will be receiving recommendations on irrigation, fertilization, and crop protection as well as yield forecasts on their computers, smartphones, or tablets.

On a side note, this kind of agricultural monitoring is increasingly being applied in the sector of forestry. It can be used for forest management and as a way to characterize forests as carbon sinks, aiming at minimizing climate change (this is part of the United Nations' REDD program)

AUTONOMOUS TRACTORS

In the vein of having technology do everything for us, state-of-the-art, GPS-guided, almost autonomous tractors, and other self-driving farm equipment are on their way to drive across our field without much help from humans. Besides, when we try to conceptualize anything futuristic and ideal, we tend to remove the human element from the equation. So, when it comes to farming, that conceptualization would be based on the use of artificial intelligence and robotics aiming at creating a perfectly efficient self-working farm whose only variable is the weather conditions.

Highly specialized, automated machinery is going to take care of both planting and harvesting. Driverless tractors will be autonomously spraying thousands of acres by use of a spot spray rig. The new tractors will also be equipped with numerous sensors that will collect data on plant health, soil composition, and field topography so that treatment is only applied to those parts of the field that need it. Soil sensors can, for instance, report the water and nutrient content of the soil. As for other self-driving farming machines, those will be robots that can pick fruit, pull out weeds and move plants around (uproot them and then replant them elsewhere).

Also, lastly and most importantly, let's not underestimate the fact that autonomous tractors have the potential to extend the working day to a full 24 hours. You see, upon receiving a weather alert, farmers can decide remotely where to send the machines to complete a task in the best conditions before the next weather event.

SMART SILOS

Smart silos bring us one step closer to the digital agriculture of the future. The way they work is as follows: Sensors are installed to monitor the inventory, that is, the quantity of every harvested good that has been stored. Then, that information will be electronically transmitted and stored as well as crosslinked with other farm data. That data will be used to create demand plans. This way, a seamless and continuous supply – in one word, **interoperability** – can be achieved.

The traditional telemetry options, originally designed for the industry, did not succeed in reaching those goals. Why? Among other factors, because of a farm's difficult access to a quality electrical grid, because of poor M2M coverage (if any!), because of bulk solids that are hard to measure, etc.

So, what are the solutions? Ready-to-use, high-tech, efficient devices for remote control of stocks. Solar-powered devices, so that they're autonomous from each farm's (probably inadequate) electrical grid. First, the feed surface is 3D-scanned so that we get accurate measurements. Then, the device is installed and connected to a free app on your phone and/or tablet. And that's pretty much it since the maintenance needed is almost nonexistent!

And as with all the aforementioned new technologies, silos to reduce a farm's administrative work by eliminating the hassles of order and inventory management, all while improving the plants' productivity. What's more, if the entrepreneur desires it, customers will be able to view the stock levels in real-time and adapt their orders.

Photo by <u>Mae Mu</u> on <u>Unsplash</u>

DRONES

Drones are not a new trend in the technology sector. Yet, thanks to investment and a relaxed regulatory environment, the time may have come for them to stand at the forefront. Drone technology is giving agriculture a high-tech makeover. Drones can generate field maps and deliver aerial infrared photos providing information on the health of the crops.

Here are six ways drones will be used throughout the crop cycle:

- **Soil and field analysis**: By producing precise 3D maps for early soil analysis, drones can play a role in planning seed planting and gathering data for managing irrigation and nitrogen levels.
- **Planting**: Startups have created drone-planting systems that decrease planting costs by 85%. These systems shoot pods with seeds and nutrients into the soil, providing all the nutrients necessary for growing crops.
- **Crop spraying**: Drones can scan the ground, spraying in real-time for even coverage. The result: aerial spraying is five times faster with drones than traditional machinery.
- **Crop monitoring**: Inefficient crop monitoring is a huge obstacle. With drones, time-series animations can show the development of a crop and reveal production inefficiencies, enabling better management.
- **Irrigation**: Sensor drones can identify which parts of a field are dry or need improvement.
- **Health assessment**: By scanning a crop using both visible and near-infrared light, drone-carried devices can help track changes in plants and indicate their health – and alert farmers to disease.

GM FOODS

The concept of genetically modifying food is nothing new. Back in the 1980s, we "re-engineered" the DNA of plants to render them disease-resistant. And by the '90s, genetically modified foods were available in our markets. Several food items that we consume today – fruits, crops, livestock, even fish – have undergone some sort of genetic modification. Crops will be made more resistant to pests and viruses, but foods will look the same as it does today. These foods are generally safe and went through strict standards. As it appears, our reliance on genetic engineering will continue to increase as we strive to feed a growing, hungry world.

New molecular biology methods make it possible to rewrite or change individual DNA blocks in genetic material – safer and more precisely than before. Opening up new possibilities is the very promising gene-editing technique **CRISPR**/Cas-9 that allows for precise genome alteration. CRISPR can accelerate traditional breeding and selection programs for developing new climate- and disease-resistant, higher-yielding, nutritious crops, and animals. CRISPR stands for clustered, regularly interspaced, short palindromic repeat and allows greater selectivity. It constitutes a technique that can breed crops with essential vitamins, nutrients, and minerals.

We have already used this technological tool to produce non-browning apples and non-bruising potatoes. It is a genome-editing method, already being used to work on developing drought-resistant corn, allergen-free peanuts, and mildew-resistant wheat. The CRISPR technology has great potential to improve food security as well as have a positive impact on several issues, such as animal welfare issues and animal-borne food-related diseases.

Photo by Mae Mu on Unsplash

BUSINESS EXAMPLES

Numerous **examples** of technology-driven businesses are already here:

In late 2017, the company **Hands-Free Hectare** became the first in the world to grow an arable crop remotely. More specifically, what they did was use autonomous tractors to sow and spray crops, small rovers to take soil samples, drones to monitor crop growth and an (unmanned) combine harvester to collect the crops. Since then, they have also grown and harvested a field of winter wheat and have been adding new technologies, tools, and capabilities to their line-up of robotic farming equipment.

In Asia and Africa, **Digital Green** is a technology-enabled communication system that brings needed agricultural and management practices to small farmers in their language by filming and recording successful farmers in their communities.

In Australia, **Sundrop** is a company that uses a seawater greenhouse. That is a sustainable system that combines solar energy, desalination, and agriculture to grow vegetables in any region, anywhere in the world. Drawing energy from the sun, it doesn't rely on fossil fuels. And it doesn't require land either.

In Ghana, **Hershey's CocoaLink** uses text and voice messages with cocoa industry experts and small farm producers.

Also, in India, **Catapult Design** 3D-prints tractor replacement parts, corn shellers, cart designs, prosthetic limbs, and rolling water barrels.

In Kenya, **MFarm** is a mobile app that connects Kenyan farmers with urban markets via text messaging.

In Nigeria, **EasyAppetite** is an online takeaway site that people can use to place their orders, make recommendations, and give feedback from their experiences and get other users' feedback as well. Much like the apps for ordering online, it includes many restaurants to choose from.

Faasos is an app that appears, at first sight, to be similar to all other national apps existing around the world where you can place your online order, read, and post reviews. But Faasos has a noteworthy differentiation: It allows users to make various choices from all its partner restaurants and place them in one single order.

In Dubai, **Agricel** enables soilless and low water growing in both indoor and outdoor conditions by using a revolutionary film farming system.

In Israel, **Seedo** is shipping worldwide its fully automated hydroponic grow system for vegetables and herbs. They have developed a smartphone-controlled, automated home-grow system. Faithful to the combat against climate change, this type of pesticide-free and space-efficient containerized farm is both water- and a power-saver.

In the Netherlands, **MosaMeat** is a startup company that uses CRISPR technology to grow meat. Backed by Bill Gates, this company intends to bring lab-grown meat to the market in 2021.

Apps such as **Calorie Counter Pro**, **HAPIcoach**, and **Noom Coach**, let us log our meals and choose when to have our next meal to follow our diet goals.

The **HAPIfork** is an electronic fork programmed to alert us when we're eating too fast.

Liftware helps people with hand tremor by neutralizing it so that eating with a fork or spoon unbecomes the tedious task that it was until today.

Biozoon prints out gourmet-looking food for seniors who need to eat purified meals.

Foodini allows us to personalize our meals with a 3D printer at home.

Soylent is the answer to those who think that eating is an unnecessary burden in their lives. It's a meal supplement powder mixed with water that contains all the nutrition necessary for an adult.

Israeli-based **DayTwo** and US-based startup **Viome** harness microbiome data for personalized diets. Both companies pull their data from different microbiome databases, but the results are promising.

InsideTracker (US) is focused on measuring the 40 biomarkers in our blood that its scientists have determined to provide the best indicators of our overall health. It gives us a personal plan to modulate said indicators with diet, supplements, exercise, and lifestyle changes.

Nutrigenomix is a start-up from the University of Toronto (Canada), available in 22 countries, delivering personalized nutrition advice and physical activity recommendations based on our genetic profile.

DNAFit and **FitnessGenes** (both from the UK) are interested in our genetic makeup to determine the most effective personalized workout. They also provide nutritional advice to help us achieve our goals and any supplements that might be of help.

John Deere has created the AutoTrac which enables huge pieces of machinery to plant crops with uniformity and accurateness. An aim is to reduce any overlap in agricultural processes such as tilling, planting, and fertilizing, which will, in turn, reduce the use of chemicals as well as increase the company's overall productivity.

Cainthus is a machine vision company that has adopted a different approach to the use of AI. They have created a facial recognition system that identifies the faces of cows in just six seconds. Their smart camera system collects video data on-site and uses it to assess animal behavior as well as any environmental changes that may affect their precision farming production. So, huge herds can be monitored easily and from afar, with minimal human involvement.

Photo by Brooke Lark on Unsplash

FOOD TRENDS

O cean life is facing mass extinction because of overfishing, pollution, seabed mining, and the destruction of habitats like coral reefs due to climate change. Yes, humans are to blame. Ironically, 3.5 billion humans today depend on the ocean for their primary food source. That figure will double in 20 years.

Fortunately, humans are aware of this and have implemented sustainable commercial fishing practices and turned to cultivate fish. Aquaculture is going big, with 35 countries producing more farmed fish than fish caught in the wild. A milestone was reached in 2011 when for the first time, more fish were farmed than beef, a trend that has continued.

Now, in more positive and exciting news, in February 2018, the French Embassy of Greece and the French Institute of Greece organized an event where Greek and French innovators would have the opportunity to present their futuristic nutritional ideas.

An entirely new product, befalling in the category of smoothies, has whey and avocado juice as its main ingredients. Whey is the spare liquid after the making of cheese or strained yogurt; it contains much protein and is used in the production of supplements for those who do bodybuilding. Yet, for many small production units, whey can be problematic, let alone a source of pollution. But back to our business. The new product in question was named Avoyog. It is not artificially colored; it contains plant fiber, prebiotic and antioxidant substances, vitamins, the protein, of course, natural antioxidants and as for sugars, just those of the avocado.

Then, essential oils were presented that were made of organic cultures of mainly rosemary and oregano. These can replace nutritional conservatives that have a blameworthy effect on our health. They can be added in a wide array of food products and make them less unhealthy; those include mayonnaise, cold cuts, tomato juices, and potato chips.

Another new product came to life because of the endless modern-day work timetables. It's called Feed and is a "bar and drink" combo that working people can consume in a day at work when they don't have time to go to a restaurant or order something. It is one of those "without" products, one that does not contain lactose, gluten, allergenic nuts, or genetically modified ingredients. It managed to collect half a mil euros of funding, so people who are interested in the future of food believe in this product.

Now, after their meal, many people enjoy a cup of coffee, so at the end of the day, the bins at the workplace are filled with a mountain of plastic cups. Nicolas Richardot, founder of Tassiopée, suggests a cup made of biscuit that keeps its temperature high for a long time, doesn't melt before at least half an hour, and finally can be eaten after you've finished your hot coffee or tea!

When FAO predicted that by 2050, we would face a serious lack of basic foods, Clément Scellier took it seriously and created a series of food products, mostly bars and pasta. He created bars with apple, cinnamon, and "flour" made of insects roasted and then grated. Those who tasted them exclaimed that they felt no difference between them and other bars that they had bought from the supermarket!

A TASTE OF UMAMI

Concerning food, the problem of the future is a complex one: We must shape a judicious, healthy, and environmentally friendly diet. In this chapter, we will see among other things, how experts have established and are certain that one of the new introductions in our dietary universe will be insects; we will be eating all kinds of insects in the not-so-distant future. So, I'm just saying it here to pave the way for what is about to follow. You might have assumed that a book about the future of food will be an easy read, but the truth is that you will need to open your mind. I mean, if you are having a hard time imagining yourself eating a scallop, how am I going to talk to you (or, how are you going to read) about cricket- or beetle-eating? It is going to be a little hard to... digest – pun intended.

But before naming the food of the future, let's do a layover in the ever-evolving land of tastes. Yes, as strange as it may sound, even the flavors tasted by humans have been evolving through the years. So, let's go back to ancient Greece for a minute. Because even in the subject of taste, my ancestors had one thing to say or two. Ancient Greek philosophers, like Plato, identified six tastes: bitter, sweet, salty, sour, astringent, and pungent. And, mind you, that was only instinctive and perceiving; there was no anatomical basis yet, that is, the taste bud within the human tongue.

Now, I just want to point out the fact that six different tastes were named back then, when even today, the basic tastes are only four: specifically, the first four (bitter, sweet, salty, and sour). Personally, I find that amazing. However, much talk has been done about two new basic tastes, the umami, and the kokumi, but we'll get to those in a minute. You see, one thing that remains a constant is people's need to lexicalize each and everything that they experience. And another thing that remains a constant is the words' power over our feelings.

On a similar note, the more we talk about something, the more it is normalized in our minds, and the more familiar we become with it, either consciously or just unconsciously. And the food is no exception to that rule. I mean, even as I read about the foods that will supposedly be massively consumed in the future, I found myself embracing these ideas more and more.

Language really affects us in ways we cannot fully realize or perceive. So, in recent years, a fifth (basic) taste has entered the world of gastronomy and our gustatory lexicon. That taste is called umami. What is umami, you might ask? Well, in the words of chef and author Elizabeth Falkner, "[...] it's quite difficult to describe umami's brothy, sea-urchin, the aged-steak flavor in a way that does it justice."

Furthermore, a possible sixth taste will be kokumi. In Japanese, it literally translates to 'rich taste' and, by extension, a taste amplifier. Isn't all that exciting?

(G)ASTRONOMY

Experts support that the diet of the future will, among other interesting things, include NASA-inspired superfood bars, 3D-printed custom-designed menus, and... plenty of kale.

Although there may be enough food to go around in the West, experts say the realities of agriculture and economics will convince more of us to become vegetarians or vegans. But if the new trends are to become more and more meatless, then plant-based protein will be asked to play a central role in the food industry's future.

According to the Food and Agriculture Organization (FAO) statistical data, protein consumption (both animal- and plant-based protein) is constantly growing around the globe. At the moment, plant-based protein is growing a little faster than animal-based, and it will probably continue to do so.

There will be a focus on foods that animals eat – since that is a reflection of what we ultimately eat. Also, there will be a focus on how technology can be used to provide us with safe, healthy, and sustainable produce.

Now, there is cultured meat, and there's also cultured fish. In fact, the biggest lab of all (NASA) gave a university scientist a grant to come up with some sort of space cuisine. The ingredients: chunks of goldfish muscle from live fish soaked in the blood of unborn calves. The result: fish chunks grew by 14% and looked like small fillets. No word if it reached the astronauts' menu. In 2015, a Silicon Valley biotech startup developed a lab-grown shrimp. However, it is not made from shrimp, but from algae, one of many recent cultured foods that have cropped up. (Algae are further discussed later.)

BIOMIMETICS

Biomimetics is the act of man imitating nature. That is, people observe an interesting structure in nature and construct something based on the same principles. In the case of researcher and Associate professor in Paris Sud, Paris-Saclay University, Raphaël Haumont, the imitation's subjects were grape and a tomato. These two fruits have a particularly thin skin, yet their content, which is about 90% liquid, does not leak that easily. Along with his lab team, he managed to create a very thin yet very durable algae-based membrane. They give it a spherical shape and can put inside it anything fluid, from water to yogurt. This membrane keeps its shape until it empties, and then, contrary to plastic cups or bottles, it can easily and quickly disintegrate, becoming one with the soil. This has even been tested in outer space without gravity, in the Airbus A310 Zero-G, a property of France's Centre National d'Etudes Spatiales.

So, for the future, we have already found a way to transport foods on this planet and beyond! Astronauts can easily pierce that membrane with a straw and drink the content without having thousands of useless small suspending droplets in the air. Raphaël Haumont makes marmalade by smashing an orange and using it in its entirety, including the white part that is rich in pectin. Adding mineral water rich in calcium, the consequent reaction results in a marmalade without a grain of added sugar but with an exceptionally rich taste and intense aroma. So, technically, we can't call it "marmalade," since it does not consist of 50% of sugar.

Haumont also talks about the frigidity of outer space that is close to liquid nitrogen's temperature, as he thinks that the latter will be increasingly used in the kitchen of the future. Drawing inspiration from how the universe is dilated in Space's vacuity, he also makes chocolate mousse without eggs. He melts 200 grams of chocolate in a bain-marie, adds 200ml of whatever natural flavoring he wants (tea, juice or water), places the mixture in a nozzle equipped with gas ampoules, puts it in the fridge for a little while and then, pressing the handle, he can serve himself as much mousse as he wants.

In the lab of Haumont's team, a fundamental axis of research is to show that natural – mechanical methods, such as centrifugation, compression, or the use of ultrasound, allow us to achieve noteworthy textures without doubtful added chemicals. They have made very plump and beefy confections without yeast. Other delicacies that Haumont and his team have "cooked up" include ganache without cream and gluten-free sweets, while their general aim around food creations is to avoid too many calories in one dish and at the same time amaze by being as innovative as possible. And they have a general prioritization of physics over chemistry.

Harald Lemke, a German professor of Philosophy at the University of Hamburg, has repeated in his interviews numerous times that what we eat is not a personal issue. What we eat is a political action and has global consequences. Right off the bat, an example would be the tuna fishing in the Mediterranean Sea. This fish is about to be extinct in this sea. Tuna fish are captured, then led to sea farms in Morocco, overfed there to get fat, then shot in the head with regular guns and loaded on airplanes to go to Japan where they constitute a coveted delicacy. So, we are talking about a real butterfly effect here!

Lemke has also coined the futuristic term "gastrosophie," combining gastronomy and philosophy. (If you're wondering about the -i.e., at the end of the word, that's the French ending.) He argumentatively persists that our eating habits are connected to a philosophy about how we want to live our life. He blamed meat as the type of food that brings about a global disturbance. Later, he proceeded to ask for meats to be handled like cigarettes. That is, to place indirect limitations to meat consumption through higher taxes and maybe even limiting its advertisements. From the suffering of animals in slaughterhouses to the destruction of forests and the health problems caused primarily by the overconsumption of meat, how can you blame someone for blaming that food category?

ALTERNATIVES TO MEAT AND FISH

Our dietary habits will need to change. A lot. And even though it sounds like a tough job, it can happen very easily through education and the right marketing. Besides, major meat companies already want in on the plant-based food market:

Tyson, Perdue, Big Dairy, Impossible Burger, Big Meat, Beyond Meat, Nestlé... It is not only critical for the environment's safety but also offers opportunities for new businesses and services.

There have been great advances in plant-based foods that can satisfy the consumers' experience and perception of meat. We have come a long way from those plant-based meat alternatives that tasted like a carton. Instead of giving up the experience of eating red meat, food technology can develop plant-based products to reduce the world's per capita consumption of red meat, one of the most important goals for both the environment and human health.

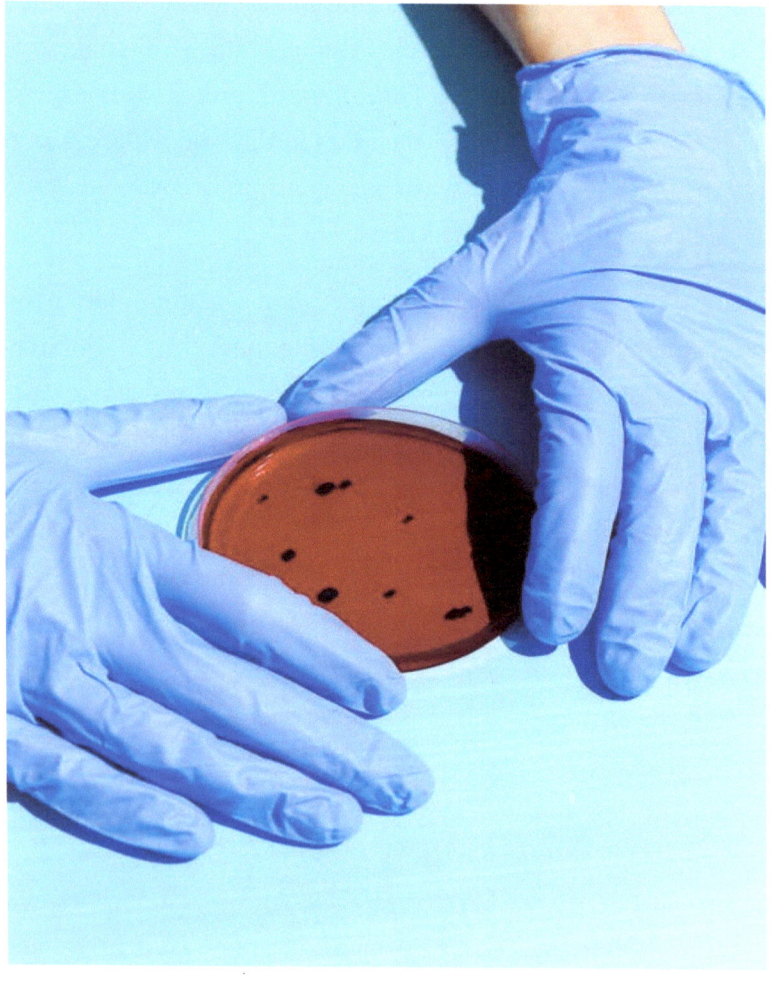

Photo by **Anna Shvets** from **Pexels**

Meat grown from cultured cells, better known as "lab-grown meat," may radically change the production of protein and food in general. The meats growing in factories include everything from fish to filet mignon. When it comes to meat, the system's jamming point is obviously the animals, and that's what innovative alternatives come to supplement, that is, the need for livestock and farms. The lab-grown animal products will include several kinds, such as pork, turkey, chicken, gelatin, milk, and egg whites.

This sounds funny, but cattle actually contribute as much greenhouse gas emissions as all the circulating cars on the planet. So, in their effort to curb global warming without taking people's craving for beef away from them, food scientists decided to develop an alternative type of meat: lab-grown **synthetic meat**. As early as 2013, scientists have already cultured ground beef from cows' stem cells. Back then, the world witnessed the first piece of lab-grown burger come to existence. The problem was that it took way too much time (two years!) and cost too much money ($300,000) to produce. Today, however, the process has become way faster and cheaper, and several "cellular agriculture" startups have already adopted such methods.

Synthetic meat can also be called cultured meat, in vitro meat, or just lab-grown meat. The process is quite simpler than one would expect. You grow muscle cells in a nutrient serum and then put them into muscle-like fibers. It's not too different a method from other 'cell culture' methods developed and used even a century ago. More specifically, we collect a small sample of muscle from live animals. This is called the 'satellite' sample, and its cells are called 'satellite' cells. Those cells are stem cells that can be turned into all the various types of cells found in animals' muscles. In theory, we could use only one cell to grow an infinite amount of meat, if we just keep feeding it with the aforementioned nutrient-rich serum!

And after synthetic meat, comes the alternative food solution that would logically follow it: **synthetic fish**. Overfishing and the pollution of oceans with chemicals are two of the many reasons why the underwater population keeps decreasing. And all that comes despite the serious warnings of ecological organizations. In fact, pregnant women nowadays are advised to avoid consuming tuna fish. Certified studies ascertain that big fish, such as tuna, salmon, and swordfish, present increased levels of toxic metals.

The business of strictly vegan alternative meat and fish is a rapidly growing one; both its demand and its development are growing. Its investments are expected to surpass six billion dollars until 2023. More and more restaurants, especially restaurant chains, create special vegetarian or strictly vegan menus.

Photo by Toa Heftiba on Unsplash

ALGAE

In a few years, algae farmed in aquaculture sites can begin to substitute feedstock and fishmeal. One of the most notable future food trends is creating products that are almost solely based on algae. While we are already using them as a biofuel, algae are seen as a solution for the problem of food shortages, since they can feed humans and animals alike. Algae is the fastest growing plant on Earth and has long been cultivated in Asia.

Plus, algae constitute a more reliable source of feedstock, since its availability is not dependent on catching fish. Food experts predict algae farming could become the world's biggest crop industry as it can be grown both in the oceans and in freshwater.

From a financial point of view, the cost of algae farming is a whopping 60% to 70% lower compared to that of fishmeal. And from a nutritional point of view, they are a good source of vitamins and minerals. Additionally, they are a favorite of vegans and wellness advocates alike. Another great thing about algae is that they grow incredibly fast, not requiring fertilizers. Considering all the advantages of algae farming, if that's not sustainable, I don't know what is!

INSECTS

Experts say that the diet of 2050 will revolve significantly less around meat and more around bugs. A tip for first-timers would be first to open your mind and then your mouth! There are about 2,000 edible insect species across the globe. Many species are non-toxic, so a new world of exorbitantly more environmentally harmless insect-based options could serve as a source of high-quality protein and micronutrients in the near future.

More precisely, a 2013 U.N. Food and Agriculture Organization report reminds us that there are 1,900 edible insect species out there that some 2 billion earthlings already regularly consume: beetles, butterflies, moths, bees, wasps, ants, grasshoppers, crickets, and locusts. Insects are abundantly available and rich in low-fat protein, fiber, and minerals.

Entomophagy (that is, the consumption of insects as food by humans) is no longer just a suggestion or an idea but real action. And it concerns not only parts of Africa, Asia, and Latin America but the entire world! Legislatures about the issue have already been established and adopted. Switzerland was one of the first European countries that formed such a law in 2017. The European Union followed in 2018.

Furthermore, let's note that **insects can (and will) be consumed in various forms**. We can eat them whole, ground, or paste, or we can extract their protein and inject it in other foods. And just like meat, insects can be roasted, fried, or boiled. For what it's worth, all of the options above are already happening at one place in the world or another.

In tropical countries, insects are often consumed whole. Now, if an insect's physiology demands it, certain body parts, such as their shell, wings, or legs, are removed. That is the case of grasshoppers and locusts, for example. And – once more – just like we mince or grind meat, we can grind edible insects into paste or powder to make them more... palatable. Then, in that powdery form, it is easy to add them to other foods and increase their nutritional value. In Thailand, they add crushed and ground giant water-bugs to a chili paste.

Thirdly, the protein extraction is a little more requiring method to apply. At first, extensive knowledge of the extracted proteins is required. Per the FAO report, "These properties include, among others, amino acid profile, thermal stability, solubility, gelling, foaming and emulsifying capacity." Plus, the cost of protein extraction is extremely high, almost prohibitive, so further research is required to make it more profitable and applicable for industry use. Lastly, I want to point out that the latter two versions (ground and extracted proteins) will render the edible insects more acceptable within societies where consumers are not used to eating whole insects.

The **nutritious aspect of edible insects** is not even an issue. They contain high-quality protein, good fats, as well as a plethora of amino acids and vitamins, totally appropriate for humans. According to data from the Food and Agriculture Organization of the United Nations (FAO), "crickets need six times less feed than cattle, four times less than sheep, and twice less than pigs and broiler chickens to produce the same amount of protein."[14]

In addition, insect breeding emits significantly fewer greenhouse gases and ammonia than animal livestock. Insects can even be grown on organic waste. And **we can use edible insects in multiple ways**. That is, we can consume them ourselves directly, we can add them in recomposed foods, and we can use them either on their own or as an ingredient into feedstock mixtures, given their valuable protein content.

The **insect conservatories** are already becoming more and more in France and the Netherlands. They are mostly for worms who love to live in the flour. In February 2018, a young French entrepreneur brought to my native Greece his famous protein bars with apple and cinnamon whose base is none other than ground "flour" of insects.

And don't be tricked into thinking that the use of insects ends here. We can also **use the fat (e.g., oil) extracted from insects**. In fact, removing fat during the production process of insect food-products prevents fatty acids (which are mostly of the unsaturated kind) from getting exposed to undesired oxidation. It can then be used to fry meat and other food products. For instance, winged termites can be fried in their own fat.

I found very interesting an international study led by La Trobe University and the University of Pennsylvania, which concluded that "people who frequently consume sushi are more open to introducing edible insects into their diets."[15] As a sushi-eater who would definitely try out an insect-based meal, I found this study intriguing. And after asking a sushi-loving co-worker of mine, I verified even further the study's results!

---

[14] Insects for Food and Feed, http://www.fao.org/edible-insects/en/.
[15] Would You Like Bugs with That? Sushi Lovers More Likely to Eat Insects, https://www.nzherald.co.nz/the-country/news/article.cfm?c_id=16&objectid=12211762.

Since 2014, FAO has published an exhaustive report titled "Edible Insects: Future prospects for food and feed security." Among other things, it includes several insects that are considered harmless, safe for consumption by humans. Some of those are the following:

- The *Imbrasia Belina* **caterpillars**: They are boiled and dried under the sun. They contain sodium, potassium, phosphorus, magnesium, zinc, copper, and iron. Concerning the latter, in 100 grams of caterpillar, there is 31 mg of iron, while the same amount of beef only contains six milligrams of iron!
- **Termites**: They contain an impressive 38% protein, plus iron, calcium, fats, and amino acids.
- **Grasshoppers**: It is recommended to collect grasshoppers for food instead of spraying on the fields.
- **Stink bugs**: Yes, even stink bugs will find their way onto our plates! Those are boiled after having been decapitated, as their repugnant liquid-weapon is in their head.
- **Worms** (in flour): Along with crickets, worms are a recommendation for the European continent, as they can stand its climate. They are rich in copper, sodium, potassium, zinc, selenium, protein and polyunsaturated.
- **Humans can eat cricket**s, but they are also reared as feed for pets. Along with mealworms, they are primarily reared as pet food in Europe, North America and parts of Asia.

Photo by Mae Mu on Unsplash

BUSINESSES WITH NEW FOOD TRENDS

In the context of promoting strictly vegetarian dietary patterns, **Good Catch** created in the lab, a tuna substitute consisting entirely of vegetarian ingredients. That substitute is made of a rich mixture comprising dry beans, chickpeas, lentils, green peas, soy, fava, spices, and finally, algae for an extra flavor and the notorious and notoriously healthy Omega-3 fatty acids.

Impossible Foods, the company-creator of the **Impossible Burger**, is now planning the creation of corresponding fish-substituting foods using the same technology is used for the Burger, which is made exclusively of plant-based ingredients yet smells and tastes just like the conventional animal-based burger we all know. The company uses a heme-extracting patent. Heme, a main component of the Impossible Burger, is a chemical element from the roots of soy plants that contain iron. On a side note, tuna fish contains heme, too. The company's founder, biochemist Pat Brown, has made clear that the company aims at having totally erased every animal element by 2035.

Founded by Ethan Brown in 2009 and based in Los Angeles, **Beyond Meat** has created plant-based burger patties with pea protein isolate, expeller-pressed canola oil and refined coconut oil as their main ingredients.

Ocean Hugger in New York has created an eggplant-based eel alternative and a carrot-based salmon alternative.

Using cellular agriculture technologies, the food startup **Finless Foods** plans to replicate fish fillets, beginning with bluefin tuna, a fish species under threat because of predatory fishing practices.

Headquartered in Silicon Valley, **Memphis Meats** is a new pioneering clean-meat company producing alternative beef and poultry. With the use of lab stem cells, either embryonic or adult ones, or myoblasts in a suitable environment of protein along with conservatives, such as benzoic sodium, it is possible – at least, in theory, to produce large quantities at 1% of the original cost. Memphis Meats easily managed to amass USD 22 million from financiers for the production of such meat. Apparently, many people can see that product thriving in the future. Among them are Bill Gates, Richard Branson, Jack, and Suzy Welch, as well as the VC fund DFJ and the food conglomerate Cargill.

THE FUTURE OF OUR EATING HABITS

Alongside the food technology and future food trends, something else that has also changed radically is the way we perceive and thus consume food. For a long time, food was seen as a clearly defined part of our lives, independent of all others. The time of eating started with our first bite and ended with our last one. Once the chicken and the baked potatoes were finished, once the sandwich was eaten, the act of eating was complete, and people's focus turned to another subject. Nowadays, the conversation about food is everywhere and can take place at any hour of the day!

Wellness is a predominant modern-day theme, and food has become an integral part of that also – as it should, by the way! It's not just mental health and practices anymore. The pursuit of our wellbeing, fulfillment, and happiness today is not about feeling good and being happily married and professionally successful. Wellness is seen more holistically than ever before, and when we talk about **holistic approaches**, we can't exclude nutrition.

Eating has become an experience, much like food production and creation have become an experiment.

Ubiquitous in social media, food has become a part of our culture. Taking neat, well-lit, glistening pictures of our food before we even have a bite has become so prevalent that it gave birth to the term **"food porn."** Our dietary habits have become part of our identity, of our lifestyle, of our health status, even of our moral compass.

In today's society, if you tell me what you eat, I can tell you what you are. At the same time, the dietary habits of consumers in industrialized countries necessitate large amounts of resources and cause climate-damaging greenhouse gases. Consequently, diversifying in the food business now more than ever constitutes an investment opportunity. Those two trends, in particular, are the ones that stand out: food is strongly associated with health and high-tech.

Eating healthily is not just an eating choice but one of our overall lifestyles and way of being. Becoming the best version of yourself is a solid, much-talked-about life goal. More and more people want to know exactly how the human body functions and digests in order to consume the necessary amounts of each nutrient. Books about diets, recipes and in general, about food, are becoming bestsellers at a fast pace. Chefs and food bloggers or vloggers are becoming highly paid celebrities and host TV shows with easily found sponsors (i.e., food companies).

And **high-tech is now shaping our lives**. Through the numerous technology tools and gadgets, the Internet has pervaded everywhere, the sector of food not excluded. Digital technologies allow consumers to have more and more influence on the value of foods. The digital information on product channels and produce content that is provided to the consumers presents a real challenge for food companies who need to adapt quickly to the new reality.

This second trend is that food's association with high-tech also relates to products themselves. Consumers have already turned to alternative sources of protein, such as soy and nuts; they have already discovered the good taste and wholesomeness of coconut milk and almond milk, and they have already recognized the cleanliness of reducing your red meat intake and preferring imitation burgers made from plant fiber.

A more direct response to the serious issue of malnutrition would seem to be to enhance, through fortification, the nutritional content of the food products that most people will probably consume. But, let's not be mistaken: The fortified-diet response is founded on an only partial understanding of how nutrition works and how the various micronutrients interact inside the human body. Therefore, such diet risks being nutritionally inadequate and, thus, dangerously unhealthy.

A cooking trend is the creation of restaurant-grade dishes at home. Consumers don't just want to be stuffed nowadays; they want to taste distinctive flavors, and they value freshness. Freshness is associated with healthy eating. At the same time, the amount spent at grocery stores has decreased due to people preferring eating at restaurants, takeaway, or delivery. Year by year, the market sees a big shift that can be epitomized as the fact that prepared meals are steadily overtaking cooking at home. And, as already stated, cooking at home is becoming more and more restaurant-like. But what happens with those who do not cook at home, those who depend entirely on the quality of prepared meals or of meals cooked by others?

Another food trend is the use of technology to turn a natural substance derived from silk into powder. One might ask, What for? Well, if you mix that powder with water and spray it on produce, "it increases the shelf life by two to three times throughout the supply chain without the need for refrigeration or any humidity control," as Andrew I've, M.B.A. '97, managing director of Food-X, reports.

The food market chain is changing in its entirety, from the way food is produced to the way it is purchased. Everybody knows that people do not want to spend too much money when buying food. But something that only a few realize is that people do not want to spend too much time either when they're grocery shopping. This unwillingness to invest any time in buying the best product for your health must be taken into consideration when businesses, large- or small-scale, advertise and promote their food products.

A 2013 *Food Technology* magazine article reported that consumers have a clearer and clearer preference for labels like **homemade, artisanal, authentic, made from scratch,** and other similar ones, while other food-products-related terms that seem to attract consumers are: **organic, farm-raised, grass-fed, free-range**, and **cage-free**. They also seem to be directing themselves towards tangy, smoky, herbal, sour, and bitter **flavors**. Thus, the purchase of **seasonings** has been multiplied in recent years. If I had to add something to those lists, that would be frozen food products, especially vegetables or vegetable-based meals. They are not to be neglected nor underestimated, as they often contain foods captured at peak ripeness.

So, all in all, the food market has seen many shifts, shifts that reflect people's lifestyles. Today's health problems, such as diabetes and obesity, are the result of excessive fast-food and packaged food consumption. Little by little, people begin to realize and recognize the consequences of one's food choices, turning towards house-made meals once again.

Nevertheless, the fast-food mania that is still going strong has dissociated people from their kitchens, and many have never acquired adequate cooking skills to prepare a more-than-decent home meal. So, food manufacturers had to find a way to meet that need, and they did: They developed food products that are already pre-cooked or semi-cooked to make it easier for everyone.

Also, more people are getting used to eating breakfast at home instead of spending money on prepackaged sandwiches and other low- or questionable-quality cafeteria food choices. This trend has influenced the products we now find on the supermarket shelves.

Photo by Amy Shamblen on Unsplash

PERSONALIZED NUTRITION: A STEP TOWARDS WELLNESS

Aloneness. Individuality. Personalization. The trends of today's societies both focus on and are shaped by perceiving people as individuals. Not as part of a group, not as part of a family, but as individuals. We have realized that wellness is a personal state, and the experience of food and eating is shifting toward that direction for both our physical and mental health. Whether it's our programmed television shows or our social media profiles, whether it's online shopping and advertising or our workout routine, whether it's medicine or the music playlists on our gadgets, every part of our daily lives is personalized nowadays. Next stop? Well, as you can guess, our nutrition. That includes what we eat, how we eat it, and how we experience it.

> *"We all have a role in the future of food, and it starts with the mouth."*
> – Dr. Irwin Adam Eydelnant, founder of Future Food Studio

After modifying our feed page on all the social media where we have created a profile, we'll be able to modify, to personalize what we want to eat, what we want our meals to contain or not. We only follow people (celebrities or not) whom we like and consciously select only to receive the kinds of news that will please us. Similarly, we'll be able to select foods that match our dietary preferences and tastes. **Gluten-free, lactose-free, vegan, raw, high-protein, low-carb, no-added-sugars, no-fats, no-conservatives, farm-to-table, organically grown, fresh, local, fresh AND local, wholemeal, multigrain**... Had you noticed how many new food labels there have been added in recent years to even the simplest food products on the supermarket shelves?

Food companies are met today with the impossible task of pleasing everybody all the time. Just like our social-media culture, our food culture has also taken a dramatic turn toward personalization. Each consumer has their ideas about what the best diet is, how each food must be cooked, how each animal must be fed, how each vegetable must be grown, all of it based on disperse pieces of information found on webpages left and right. So, food companies need to be all-inclusive to survive in today's misinformation or over-information or half-information.

For instance, pizza parlors would dispose of a single type of dough in the eighties and nineties, but now they must provide a variety of pizza doughs for the consumer to choose from. Some of those choices might be flatbread, cornbread, thin crust, thick crust, cheese-stuffed crust, Sicilian style, Chicago deep dish, Neapolitan crust, pan crust, focaccia, and calzone. Are you hungry yet? I know I am.

All this perfectly falls in line with the Internet's hard-to-beat fast pace. When you are used to reaching any information within a matter of seconds, you like it and want it to extend throughout your life. So, it is only natural for a generation that has grown up with the instant informing, and gratification that Google can provide to now want equally instant gratification when it comes to food. The consumers of 2020 want both speed and specificity. And they don't hesitate around it.

With easy access to all cuisines of the world through food-delivery applications, you can have any type of food you desire in an instant. Instant food ordering. Instant photo uploading. Instant music downloading. Instant noodles. Capriciousness is taken as a given in the market, and entrepreneurs have translated it into a commercial opportunity; they try to satisfy every quirk and gain as many new customers (or clients) as possible.

The new goal is for each person to enjoy the most pleasure out of their eating experience daily. The days when the mother was in charge of food and cooking when you had to eat whatever she had prepared for the family, regardless of how much you liked it or not, and be thankful for it as well, seem to be long gone, a thing of an ancient era. Today, each product enters the market in many different variations, one for each palate. For example, we can find white rice with peas and corn, yellow rice with mushrooms and beans, brown rice with corn, chickpeas and raisins, parboiled rice with peppers and onions, all of them frozen and requiring no more than fifteen minutes of boiling.

Together with what we eat, how we eat it, cook it, produce it, and distribute it, another thing that has seen dramatic change is with whom we eat. More and more people (mostly adults, but also children) have been eating by themselves this past decade, and that number will probably be increased in the near future. Eating home-cooked food together as a family has become less common.

We have developed a taste for delivery or takeaway that has been slowly but steadily killing off the heart of the house, aka the kitchen. As fewer and fewer people are willing to cook, the cooking room is about to experience a radical change in the future, just like the entire food business chain. Instead of having an entire room dedicated to the kitchen, we might have a little corner with just the bare essentials of a kitchen, that is, a fridge, a microwave, some cutlery, and a few plates and glasses.

As a consequence of all those changes, the food delivery market had to change and adapt to the new reality. New apps with more and more features, faster speeds, and increasingly competitive deals are just a few of the freshly established staples for the future of food delivery.

California-based startup **Habit** sends us our personalized meal plan based on our genetic markers. Another company, **Thryve**, allows us to determine the composition of bacteria in our digestive system (i.e., our microbiome, which we've already met) and thus essentially define our diet. In the future, our smart refrigerator could alert us concerning the personalized foods we'll need to buy each day. Also, more food delivery services will specialize in catering to our exact needs. We could be walking into restaurants, provide a saliva swab, and be served a meal that's designed to our personalized nutritional needs.

THE FUTURE OF FOOD DELIVERY

In 2019, one can easily say that even food is on its way to becoming an online business. Boosted by cellphone applications and decreasing delivery fees, delivery is the new way of ordering food. Ordering in is part of the newer generations' reality, while the generations who cook daily are about to eclipse. So, the consumer demand for ordering food online is rapidly increasing.

The new platforms for placing food-delivery orders allow consumers to compare prices, menu varieties, read and write reviews as well as order very fast, even with a single click in case they want to place the same order as another time.

Those who join this new kind of delivery must do everything to meet the needs of the customers of the future. When it comes to food delivery, most orders are placed from home on the weekends, and the most important aspect is the speed of delivery. Another part that needs to be taken into consideration is that once someone has signed up to a food delivery platform, they will most likely stick to that one platform and not change it unless there is great disappointment one or two times.

Fast-food chains and restaurants alike now need to make delivery part of their strategy and marketing, as they are doing. As estimated by Morgan Stanley's research, 40% of total restaurant sales – which translates to USD 220 billion – "could be up for grabs by 2020." At the time (July 2017), the respective current sales stood at about USD 30 billion. Consumer interest in food delivery is growing by the year and thus is now not just a marketing focus but a key investor debate as well.[16]

By far, the most common form of delivery is the traditional model, in which the consumer places an order with the local pizza parlor or Chinese restaurant (although many other kinds of restaurants, particularly in urban areas, now offer delivery) and waits for the restaurant to bring the food to their door. This traditional category has a 90 percent market share, and most of those orders – almost three-quarters – are still placed by phone.

However, as in so many other sectors, the rise of digital technology is reshaping the market. Consumers accustomed to shopping online through apps or websites, with maximum convenience and transparency, increasingly expect the same experience when it comes to ordering dinner. Online delivery times have gotten shorter and shorter, as retailers move closer than ever to their customers' most personalized items.

The business of delivering restaurant meals to the home is undergoing rapid change as new online platforms race to capture markets and customers across the Americas, Asia, Europe, and the Middle East. Although these new Internet platforms are attracting considerable investment and high valuations – already, five are valued at more than $1 billion – little real knowledge about market dynamics, growth potential, or customer behavior exists.

---

[16] Alexa, What's for Dinner Tonight?, https://www.morganstanley.com/ideas/online-food-delivery-market-expands

That growth has already taken place in Europe. The online ordering rates stand at 56% in Sweden and at 43% in Austria, to name but a few. It is no wonder that the most massive growths are expected to occur in regions where smartphones also have dominated the market. As of now, technology perseverance and online delivery perseverance are not necessarily associated, since the food players have managed to expand geographically. But in the future, the smartphone market will denote potentiality for the growth of the food delivery market.

Worldwide, the market for food delivery stands at €83 billion, or 1 percent of the total food market and 4 percent of food sold through restaurants and fast-food chains. It has already matured in most countries, with an overall annual growth rate estimated at just 3.5 percent for the next five years.

Restaurants have started looking beyond dedicated delivery startups to potentially working with much bigger players in retail who could offer robustious logistical infrastructure, larger operational scale, and wider consumer reach. And what could be the best example but Amazon? gg

Accordingly, online food-delivery platforms are expanding choice and convenience, allowing customers to order from a wide array of restaurants with a single tap of their mobile phone. Customers drawn to the new online food-delivery platforms have a different set of needs and expectations from the traditional pizza customer. And those online engines offer recommendations based on data provided by the customers, which for food companies is an asset that cannot be overestimated.

AGGREGATORS: THE NEW DELIVERY

Aggregators build on the traditional model for food delivery, offering access to multiple restaurants through a single online portal. By logging in to the site or the app, consumers can quickly compare menus, prices, and reviews from peers. The aggregators collect a fixed margin of the order, which is paid by the restaurant, and the restaurant handles the actual delivery. There is no additional cost to the consumer. With their asset-light model, aggregators post earnings before interest, taxes, depreciation, and amortization (EBITDA) margins of 40 to 50 percent.

These new-delivery players allow consumers to compare offerings and order meals from a group of restaurants through a single website or app. Crucially, the players in this category also provide logistics for the restaurant. This allows them to open a new segment of the restaurant market to home delivery: higher-end restaurants that traditionally did not deliver. The restaurant compensates the new-delivery players with a fixed margin of the order, as well as with a small flat fee from the customer. Despite the higher costs of maintaining delivery vehicles and drivers, the new-delivery players achieve EBITDA margins of more than 30 percent. Players include brands that operate globally, such as Deliveroo and Foodora, which are continuing to capture new regions. We believe that the addressable market for new delivery will reach more than €20 billion by 2025.

The aggregators' opportunity is to extend food delivery to a new group of restaurants and consumers. Rather than competing directly with the aggregators, new-delivery players are expanding the overall market. However, it is possible that in the future, even lower-end traditional-delivery restaurants will migrate to new delivery because they will find it more cost-efficient to outsource logistics; thus, new delivery poses at least a potential threat of disruption to the aggregators.

The aggregator business model has attracted significant investment around the world, allowing aggregators to advertise widely and build recognition for their brands quickly. **GrubHub** and **Just Eat**, for example, each reported marketing budget of about €70 million in 2015. Since there is no limit to the number of restaurants these platforms can sign up, once they enter a market, they can grow rapidly.

Photo by FOODISM360 on Unsplash

MEET CLOUD KITCHENS

What do you get if you combine the dominance of technology – in this case, the advent of food delivery apps – with people's unwillingness to go out to restaurants and they prefer to eat at home? Cloud Kitchens. No, they do not operate up in the sky. Cloud Kitchen is the concept of a restaurant having an area, a base, where food is cooked and packaged and working on delivery mode alone to respond to food orders that were placed online.

The new concept of Cloud Kitchens is a very promising, realistic, and viable business model. You only need a large kitchen and adequate storing space but no chairs-and-tables areas or gardens. So, you have a smaller business place and pay a lower rent than a conventional restaurant. Plus, you'll have lower infrastructure costs in the beginning and lower maintenance and cleaning costs after that. This makes it an easier, more attainable startup venture for new entrepreneurs. Also, you only need cooks and deliverers but no waiters or cashiers. So, you save a lot of staff salaries.

Then, Cloud Kitchens also have the following two options, which is rarely ever the case for other business models within the food and restoration industry. They can either tie-up with a food aggregator (see right above) to deliver food, or they can hire their delivery team. And this applies regardless of the number of locations where such a "cloud business" is expanded.

As mentioned, Cloud Kitchens work with online food orders. Having your order-online system and/or a mobile app allows you to collect pivotal customer data that helps in further engagement with your customers as well as planning and developing newer and improved, targeted, customized menus. You have more freedom to explore and experiment with your menu. In brief, you can easily – to some extent, automatically even – collect customer feedback to improve the business's overall performance.

THE FUTURE'S KITCHEN: A WELLNESS-CENTERED ROOM

Focusing on all the above, namely the food delivery domination, the increasing speed of life, the automation, etc., one would suggest that in the future, we might only have our 'kitchen' dispersed around the house; we might dispose of a utility room for washing and a larder for food storage. The living rooms could become bigger, as the sofa in front of the television may replace the kitchen table or even the dining room altogether. Once again, we can see how everything in life, every sector of it, is interconnected: the rise of the obesity rates, the dominance of technology along with the overuse of technological gadgets, the developed taste for personalization, and the developed distaste for spending time to prepare or create something that you can also possess instantly.

But frequently, a certain change brings about a counteractive wave to sort of balance things out. For instance, music has become primarily digital, streamed, downloaded, or bought from online retailers and streaming platforms. The change that is this digitalization of music has also caused an increase in vinyl sales, a retro, more traditional options. Now, when it comes to food, the **digitalization of food and food delivery** in a fast-pacing world has made many people turn to more organic options, such as growing their own food.

That's also what the trend of veganism demonstrates. That's also what the numbers of the USDA[17] show; of course, that's not just an American trend. In Europe, the average growth rate of the organic products' market has been on the rise since all the way back in 2006. Our distancing from Mother Nature has brought forth this instinctive worry that we need to reconnect with her. And all those trends are the counteractions of the digitalization of food.

It's like the Physics axiom that a reaction accompanies every action. We discussed earlier how the kitchen of a hard-working, no-free-time-having individual would be. Barely existent. That is the action taken to survive in a time- and money-demanding society. But at the same time, technology is creating a counteraction

---

[17] United States Department of Agriculture

which is making our kitchens better, fancier, more efficient as well as more... personal!

As the title swiftly suggests, you can call the kitchen of the future "The **Wellness Kitchen**," a term coined by Veronica Schreibeis Smith, founder, and CEO of Vera Iconica Architecture. This future trend aims at thoroughly transforming this vital room of our houses into a more accurate reflection of ourselves. And it's not going to be just about our palatal preferences. The Wellness Kitchen will nourish rather than feed us, taking care of our physical health, our soul, not leaving out the planet either.

In the future, our kitchens will be vibrant, full of life, easier to clean, simpler to use, more convenient, and even sunnier! How? Read on to find out! If someone asks what the kitchen is for, we'll probably answer that it is the room of the house where we cook and eat. Yet, other than food preparation and consumption, the functionality is far more elaborate and goes on to include food storage, cleaning, disposal, and the Wellness Kitchen is about to contain even more!

Before we move on, let's point out that the kitchen, as we know it today, developed its image after the second World War, and its many cupboards aimed at storing tons of canned foods and sacks filled with all kinds of grains, rice, and legumes. The 1950's kitchen was not meant for fresh fruit and veggies – let's make that one thing clear. So, you can see that in a post-war era, things have changed radically.

With minimalization being at the epicenter of the future kitchen's form and design, the new kitchen will have more yet much smaller storage parts, which can mean that there'll be more space for gathering and sitting, and every one, our teenage kids included, will be able to become a chef breezily.

There'll be cabinets where both temperature and humidity are controlled so that each ingredient luxuriates under its best-fitted conditions until consumed. This will also be true for the fridge. More yet, smaller shelves and drawers will optimally serve the refrigerated ingredients. And as for the fridge doors, they will be preferably glass so that you can see the healthy and full-rainbow-colored fresh produce inside and be tempted to taste all that richness before any of it perishes and goes to – literally – waste.

Exponentially less packaging, be it tin or plastic, will occupy space in our wellness-centered, environmentally friendly future kitchens. This means that no harmful chemicals will sleazily migrate from the packages to our food and then, digestive system. Also, that much less packaging will lead to a much smaller contribution to landfills. Plus, we'll have much less garbage to take out.

Moreover, as we discussed earlier, we'll be able to grow our food not only in our yards and orchards but also *inside* our very kitchens. Drawers full of growing plants will be the new hencoop. We will have "window gardens" with farm-fresh produce, almost at arm's reach from the kitchen sink!

Remember when I said that the new kitchen would be sunnier? Well, by having significantly less space dedicated to today's cupboards that currently store large packages of non-fresh food, there will be more space available for windows. That one small kitchen window that we have right now will be replaced by a big, maybe three-piece, a window that will brighten up our house's most important room and let the sunlight reach our home-grown herbs, spices, and veggies through the glass that covers them.

Now, what's a future without integrated technology? We cannot escape it, and why should we? More specifically, it will be imperative to have access to healthy, clean drinking water, which will be achieved through a water filtration system integrated with a primary faucet. Also, the oven's range hood will vent to the exterior. And we could have a compacting composter instead of a trash can. The former will freeze our food waste and kill the bacteria in it, thus mitigate any unpleasant scents, which are caused precisely by the proliferation of those bacteria.

Minimalization won't be limited to the kitchen's design and cupboards but will expand to our plates. Smaller dish sizes have been associated with less eating in a series of psychology studies. It's not by chance that the appearance of larger plates was simultaneous to the appearance of higher obesity rates. Fortunately, more and more people today become aware of that proven correlation.

Next stop: Health. Both physical and mental. Let's start with the physical one. The arrival of plants growing in our kitchen automatically means that we are going to have cleaner air in our home. Breathing cleaner air leads to thinking more clearly, feeling less tired, and functioning better as a whole from a physiological point of view. Also, the materials themselves will help keep toxicity out of not just the kitchen but the entire house. Specifically, recyclable glass, along with natural wood and stone, will take the place of the toxic plastic and aluminum.

Of course, another contribution to the clean air will be the existence of much less waste, as noted earlier. And the most obvious of the obvious is that cleaner and fresher eating is healthier eating. Oh, and the more colorful, the better, if we're going to be even more accurate.

Now, let's get to the mental health that, much like what happens with the plate sizes, more and more people learn and realize its importance in our lives. So, the ideas that will be central to the wellness kitchen are small-sized (and conditions-specific) cabinets, adjustable drawers, shelves, and other types of surfaces, all of which practically means more space to move around, more room for people to do activities.

In other words, the open-space kitchen of the future will allow all four or five members of the average family to be in the same room with ease, one tending the herbs, one putting the groceries in the fridge, one washing the lettuce, etc. And that's not a coincidence. The wellness kitchen specifically aims at promoting communality. Instead of seeing the kitchen as just the meal-preparing place, you'll be seeing it as more of a social meeting place, a part of the house where you'll enjoy being and feel good while you're in it.

Overall, the new trend that is wellness will not disregard our kitchens. After all, our nutrition is a fundamental part with regards to our wellbeing. Lastly, I'd like to point out that, even though I used the word 'future' several times, none of the above can be characterized as 'futuristic.' All those new elements are already existent and very specific in their development and aim – they're just waiting for us to become more conscious about our eating, cooking and consuming habits and how they affect both ourselves and our planet.

FOOD SHARING

Thanks to technology, we have seen many types of goods and services being shared, "passed around." This includes domiciles (houses, rooms, summer homes, or apartments) and cars. Now, it is expanding to every industry, and food is no exception.

Founded by social entrepreneurs, **Olio** has developed an app that connects people with their neighbors and local shops, aiming at sharing any surplus food instead of discarding it. It serves a smart technological response to the issue of excessive food waste.

In London, **The Cookhouse** is a bookable combination of an open-plan kitchen and dining space at Borough Market. It is based on the idea of community cooking in a communal kitchen. Seeing that we talked about the shrinkage of our kitchens earlier, the truth is that there are more important things than the size of that room, such as eating with others and taking time out to properly enjoy eating and/or making food, two of community cooking's perks.

Photo by KOBU Agency on Unsplash

THE FUTURE OF WATER

Water: For some, it's a basic staple taken for granted, while for others, it's a hard-to-access luxury. It has become a grave issue, whether because it's too much due to flooding or too little due to drought. According to the World Health Organization (WHO), approx. 844 million people are deprived of drinking water and approx. Two billion use water taken from a source contaminated with feces. In response, as the water shortages are getting worse, the United Nations realized that there was a need in this world to recognize safe drinking water and affordable sanitation as human rights.

But the problem doesn't end there. Of all the water in the entire world, only two percent of it is freshwater, and for a variety of reasons, only 25% of that is accessible to humans. So, the entire human race has survived up until this time with this minimal percentage. **Atmospheric water harvesters** – an MIT project in Peru, which is the future's miraculous solution to this vital issue – can change that. Specifically, they can suck moisture out of the air wherever and whenever needed!

Furthermore, the United Nations declared 2018 to 2028 an international decade for action on the water. In essence, the water challenges that we must face can be divided into the following two categories:

- Water resource problems, and
- Water infrastructure challenges.

All human threats are linked to water.

Whatever this planet and humanity have to face is linked to water. From natural or humanmade disasters to extreme weather and climate change, from biodiversity losses (aka species extinctions) to ecosystem collapses, everything is largely associated with the cycle of water or water contamination. Whenever lives and livelihoods are destroyed in masses, water shortage or water contamination lies at the heart of the problem. In addition to the dramatic statements, water has ranked in the Top 5 risks in the World Economic Forum's "Global Risk Report" for seven consecutive years!

So, time to get to the solutions. Much like the challenges, the sources from which the solutions are also two:

- (Local) Institutions and
- Technology.

The institutions can be either political or socioenvironmental. As the years' pass, the role of local institutions in water resource planning and management is becoming increasingly crucial and defining. As for technology, emerging technologies can supply previously unobtainable data and provide solutions to global water shortages.

> *"Water is the driving force of all nature."*
> **– Leonardo da Vinci**

A very interesting and admirable initiative is the Water Resources in the South East (or **WRSE**) that was formed in the UK in 1996. WRSE has a plain and simple goal: to develop new ways of distributing water taken from existing sources, using existing pipework. And then, that goal can be taken one step further: that is, to develop new, bigger sources of water and new, wider, and longer pipework to distribute the water across a larger region.

A world-changing project was developed by a French designer and is called the **Warka Water Towers**. It is primarily applied in Africa and has changed the lives of a lot of young African women. They've also been taught how to build such towers on their own; a team of four people can build one in a week. It gets roughly a hundred liters (approx. 26.5 gallons) a day, depending on the humidity.

Portsmouth Water and **Thames Water** are both planning a new water reservoir that they'll be able to share with their neighbors. They collaborate with Southern Water and Affinity Water, respectively. After that initiative's success, taking a regional water management approach is now being promoted as an important step in building resilience across the country.

Another business example of proper water management application is **NEWater** in Singapore. This new brand was created to introduce and promote the idea of recycling sewage water.

Sustainable growth and development just cannot take place in cities unless city wastewater is properly recycled and used as a resource. Given the rising global water shortage, cities can no longer pour away, something that is turned into a valuable resource. In the words of Rudy Rooth, principal consultant at DNV GL in policy advisory and research for sustainable energy use: "Proper water management can give value to materials previously deemed harmful or even dangerous for the environment. For example, wastewater contains a lot of nutrients – nitrogen and phosphates – that can be recovered at the wastewater treatment plants and used in fertilizer."

Additionally, wastewater treatment is energy-, chemical- and transport-intensive, so recycling it makes not only environmental but also economic sense. "According to the Organisation for Economic Co-operation and Development (OECD), the key water areas to be addressed in any integrated water management system are reduced water use, reuse, and sludge management. While rainwater harvesting, sustainable urban drainage systems, water efficiency, leakage reduction, smart metering, and water tariffs all have a role to play, wastewater treatment lies at the heart of any closed-loop water management system."[18]

To address the complex water management issue, the OECD has also launched its Principles on Water Governance, calling for sustainable, integrated, and inclusive water management in cities.

Customized irrigation management both increases crop yield (as we saw earlier) and reduces water consumption. Irrigation is agriculture's fundamental factor. Effective irrigation-water management can respond to fluctuations in the water supply. A distributed wastewater treatment system can also be used to cut the high costs of transport and distribution. So, it makes perfect economic sense. This is of extreme importance because most of the increase in agricultural production in the future will come from yield increases. After all, increasing arable land is a bit unfeasible.

Lastly, observing Earth from space can help governments manage water resources, mitigate humanmade contributions to climate change, and prepare for natural disasters before they occur. In his research titled "Emerging Trends in Global Fresh-Water Availability" and published in *Nature* in May 2018, Dr. Matthew Rodell, chief of the Hydrological Sciences Laboratory at Nasa Goddard Space Flight Center in Washington, discovered that the planet's wetland areas are getting wetter, while its dry areas are getting drier due to several reasons, including climate change, natural cycles and unbecoming human water management.

Currently, his team in the Laboratory is working on a drought-monitoring product that will provide real-time drought maps and flooding risks across the globe. However, water usage cannot be monitored from space. Hydrologists will keep depending on individuals, businesses, and states to cooperate and provide accurate data concerning their water usage. And that is always a sad and not very hopeful truth for scientists!

---

[18] Raconteur, "Treating City Wastewater as a Resource," article by Felicia Jackson, September 21, 2018, https://www.raconteur.net/sustainability/wastewater-resource-city

Photo by Jacek Dylag on Unsplash

EPILOGUE

I f the future of our food involves so many different sectors (human health, the global environment, chemistry, technology, the economy, businesses around the world, financial policies, climate change, animal rights, etc.), it's because everything is interconnected, and due to our increasingly developing civilizations, people have become the epicenter of that interconnectedness.

Now, the very reason why we're even talking about "the future of food" in that way is because the global population is constantly increasing dramatically, and the current (or conventional) agricultural systems and livestock farming are not sustainable nor as environmentally respectful as they'll need to become, while at the same time noncommunicable diseases are on the rise, with obesity, diabetes and cardiovascular diseases standing at the forefront of the global health-related issues. And all the above is worsened by the partly associated global issue that is climate change. So, as mentioned above, several sectors must rapidly evolve to cope with all the existing challenges appropriately.

The future of our fields. One obvious needed change concerns our very fields. And I list it first, as it also constitutes the first stage in the long chain of food production. So, there will be significantly fewer people and more self-driving machines on the field, accompanied by satellites, sensors, drones, and silos, among other things. With the help of technology, the field – and information about it – will come to you more than you physically visit it.

The future of our farming. There is an urgent need for farming to become land-intensive, as part of the environmental awareness. The surface of arable land is limited, so for the future, we're developing new techniques, such as vertical farming and underground farming, as well as new types of greenhouses, including in our kitchen drawers, and experimenting with different types of lights that best fit each species of vegetables, legumes, and herbs. Plus, insect-growing is about to become a new expanded form of culture.

The future of our dishes. Ingredients-wise, the main difference between then and soon is the inclusion of several insects in our dishes... but not their appearance on our plates as well. You see, in the West, insect-eating is not popular, so they'll have to be consumed in an imperceptible form, such as that of a powder. Also, the dishes of the future will be of better quality and, hopefully, of smaller quantities as well.

The future of our kitchens. When it comes to this room that we call the kitchen, a part of the house that's considered almost holy in my native Greece, I have concluded that there will be two opposite changes. After finishing my research for this book, I had gathered a lot of sources stating that fewer and fewer people like or are interested in cooking, while numerous other sources supported that future kitchens will be... well, futuristic, seeing all the new appliances being developed, the new equipment that will make the cooking process easier and more convenient. So, I believe that in the future, the number of both 'minimal' and 'maximal' kitchens will increase. The two kinds of a kitchen that you'll see more and more frequently are the following: One, a minimalistic corner with the bare essentials instead of an entire room, and two, a sophisticated, high-tech kitchen with 3D printers and herbs grown in cupboards.

Censored by censors. The development and application of sensors is a recurring concept in this book, as they will play a role in every stage of food production but also star in the last scene of the food's consumption. Censors in the forests to prevent any (illegal) felling, sensors in our fields and farms to notify us about the soil's conditions promptly, sensors in our bathrooms to tell us in real-time what we need to eat and what we should avoid. In a way, sensors will be censoring our mouth. Sensors will be the new security cameras. However, unlike the conventional censorship methods, "censorship" will not censure what comes out of our mouths but rather what goes in them! And since this will be straight to the benefit of our health, I don't think we should have any complaints whatsoever!

A.I. Combos. The food of a future nine-billion world population will largely depend on effective and efficient collaborations among new pieces of machinery and technological equipment. Artificial intelligence will take over and facilitate processes in the fields, in the farms, in the kitchens, and the food-delivery business. Information will be centralized and spread out fast. If you combine the sensors mentioned above with autonomous tractors, automated irrigation, and a multitude of cellphone apps, what you get is much more effective and much less fatiguing farming, more efficient exploitation of each square inch of arable land, and less food waste. And that combination-rich cake will have a water- and the energy-saving cherry on top.

Politico-financial co-ops. But apart from technological coordination in the course of many production processes, that same level of collaboration and cooperation is needed in the political and financial aspects of food production and across the stages of food trade. Regardless of how far on the right or the left, any given government might stand, in the future, all states and policymakers will need to support startups related to the food industry. I'm talking about both the food-producing and the food-selling companies. (The combination of the two is obviously included.)

After all, the **involvement of the private sector** will once again be of the utmost importance. If new technologies are to be adopted and put into practice successfully, it's mainly the private sector that can ensure such a thing. Also, the private sector – for instance, in the form of food producers, supermarkets, or restaurants – has a shared responsibility for assuring that they provide their customers with a range of healthy and nutritious foods.

Overall, each and every big problem or small issue discussed in this book requires close collaboration between food production units, lawmakers, food scientists, states, and environmentalists. Close co-operation is also required with research institutions that play a crucial role in transferring knowledge and ensuring the application of the best practices. Furthermore, governments must continue more strongly in the process of educating their citizens about the concept of sustainable development and all its benefits. Public awareness campaigns have been shown to influence national conversations and effectively change the statistics, such as cutting food waste in several countries and making the whole world more environmentally knowledgeable, aware, and sensitive.

Personally, I'm both optimistic and curious – opti-curious, if you want – about the future in general and especially around food. I do believe that all the initiatives aiming at the development of innovative food production methods will take place, as they have already started getting backed and financed on a large scale. And I am equally optimistic concerning the sugar taxes (or, preferably, health taxes). Seeing the health-wise beneficial changes brought forth thanks to alcohol and tobacco taxations, taxing sugar will certainly help combat the rise of obesity and diabetes rates, just like the former helped combat the rise of respiratory diseases, which can also become deadly. As for my curiosity, that mostly refers to the new kitchen equipment as well as to all the new ingredients, tastes, smells, and colors that will soon take over, beginning from the developed countries.

Photo by Caju Gomes on Unsplash

Iris Efthymiou – Egleton

Author- Speaker- Entrepreneur
https://www.irisefthymiou.com/

Iris serves on the boards of and is associated with a number of organizations involved with the economic activities of the humanity. She is an author, speaker, entrepreneur, and traveler who has published articles and written books on an eclectic mix of issues. These reflect her firm belief in the power of combining different disciplines in order to arrive at a more balanced and nuanced view of the topic concerned.

Iris has spent the last few years researching and writing about the issues that she believes we will face in the fast-changing world of the future.

She studied Economics at the University of Athens and earned her 'Masters' at the University of Piraeus, where she is now a PhDc. Together with the continual updating necessary for her business career, She has continued to maintain and widen her knowledge base by undertaking numerous external studies at, inter alia, The Karolinska Institute, Harvardx, and MITx.

She had worked as a college director, created her own Public Affairs business. She worked with politicians, ambassadors, entrepreneurs, academics, journalists, and spoken at the UN Headquarters in Geneva. In recent years, as her priorities changed for various family reasons, she has focused more on publishing a number of papers and written 10 books including her 2019 book "Practical Powerful Parenting"

Iris' other interests include being an active member of the Board of Academics and President of the Interdisciplinary Committee. {HAPSc has Special Consultative Status with ECOSOC and is a Member of United Nations Academic Impact (UNAI)}.

She is, also, a Scientific Associate of the Laboratory of Health Economics and Management (LabHEM) of the University of Piraeus.

In September 2019 she joined the Board of Womanitee – a Greek initiative to create a network of Centres for the Promotion of Female Entrepreneurs, acting under the auspices of the Ministry of Development and Investments, the Ministry of Work, and the Central Union of Municipalities in Greece.

Harry Efthymiou – Egleton

Harry Efthymiou - Egleton is a seventeen-year-old student currently studying in the UK. He has a passion for new technologies, politics, philosophy, as well as sports. He has worked in and with the Hellenic Association of Political Scientists (HAPSc*) to help widen access to education for students of all backgrounds and helped to organise events that attempt to broaden the political debate with leaders and academics from around Europe.

As well as being a genuine foodie, Harry is an avid reader who also loves debating and who enjoys Model United Nations competitions.

*{HAPSc has Special Consultative Status with ECOSOC and is a Member of United Nations Academic Impact (UNAI)}.

As you might notice, I have chosen to deal with various subjects, from politics to the environment and from medicine to diet, while my repertoire will only keep growing. Our society is like our body, and both are like the entire universe: wonderfully complex and unchangingly interconnected. I deeply believe in the interconnectivity of everything in the world as we know it. It is one thing that science, philosophy, and spirituality agree on"

Some of Iris Books

- *Trends in Healthcare, A global challenge.* \
- *Do we really Know China?*
- *Wellness, a new word for Ancient ideas*
- *Human Health and Oceans: Every second breath we take comes from our ocean.*
- *Practical Powerful Parenting: Be the parent you know that you can be!*
- *Gratitude creates Solutions.*
- *Look Inwards for Rewards: 70 questions for Self-awareness.*
- *You, Your Kids and Money.*

Most of the Books are also available in Libraries, inter alia:

- Bodleian Libraries of the University of Oxford
- Cambridge University Library
- The National Library of Scotland
- The Library of Trinity College Dublin
- National Library of Wales
- Library of Congress

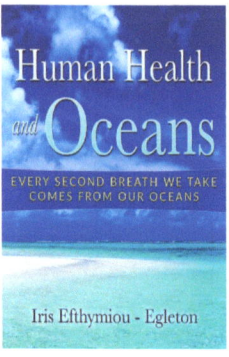

See what they say about *Iris*

"Iris, you are a true inspiration of how the woman of modern world shall look, behave, think and act. In one word you are PHILOSOPHY put in Reality. Woman of wisdom, knowledge and understanding, with a big warm heart full of balanced positive emotions...our Greek Goddess of Modern times" M.F.

"Iris is a quality asset to any kind of effort, because of her inspiration and motivation skills, elements that enhance any team's performance..." T.D.

"Iris is a person who knows what is required in any situation that she may be involved in..." N.G.

"She will always be an inspiration for me, not only as an extremely professional woman, but also as a writer and a highly creative mind with a positive attitude... I have always had her as a role model...and I am so blessed that she had helped me..." M.T.

"...You have a great experience and knowledge in human affairs which makes you a role model for many around the globe." N.S. - Diplomat

SOURCES

1. Envisioning a Future Without Food Waste and Food Poverty: Societal Challenges https://www.wageningenacademic.com/books/doi/10.3920/978-90-8686-820-9
2. Food Trends: Now and Future https://www.healthcentral.com/article/food-trends-now-and-future
3. Future Diets: The Global Rise of Obesity https://www.odi.org/opinion/9329-future-diets-global-rise-obesity
4. Menu for the Future https://www.nwei.org/discussion-course-books/menu-for-the-future/
5. Population and Food: Global Trends and Future Prospects https://www.taylorfrancis.com/books/9781134811694
6. From dystopian fiction to just-as-scary present-day science, here's your required reading to unpack some of the planet's biggest changes in the past 100 years https://www.saveur.com/required-reading-future-food-books
7. Can We Feed the World? The Future of Food https://www.scientificamerican.com/store/books/can-we-feed-the-world-the-future-of-food/
8. The Future of Food: Shaping the Global Food System to Deliver Improved Nutrition and Health http://www.worldbank.org/en/topic/agriculture/publication/the-future-of-food-shaping-the-global-food-system-to-deliver-improved-nutrition-and-health
9. The Global Food Economy – The Battle for the Future of Farming: A comprehensive guide to the issues affecting world food production. https://www.zedbooks.net/shop/book/the-global-food-economy/
10. Amaranthus: A Promising Crop of Future https://www.springer.com/gp/book/9789811014680
11. Is There Enough Meat for Everyone? https://www.gatesnotes.com/Books/Should-We-Eat-Meat
12. Future Food Is Artificial https://tvtropes.org/pmwiki/pmwiki.php/Main/FutureFoodIsArtificial

13. Almost nine years of The Food Futurist and a new approach
    https://hfgfoodfuturist.com/
14. Protein crops: Food and feed for the future
    https://www.frontiersin.org/articles/10.3389/fpls.2017.00105/full
15. As the Pool of Agribusiness Giants Shrinks, Will Innovation Follow?
    https://worldview.stratfor.com/article/pool-agribusiness-giants-shrinks-will-innovation-follow?utm_campaign=LL_Content_Digest&utm_source=hs_email&utm_medium=email&utm_content=34015433&_hsenc=p2ANqtz-8-n2C6srviFjgh0r0s0YJ8VLWFyl19-YtzBABH40BoKtjAEs7cUnBegREGI8bwIoDN_fv2RhdoM1yyvWruXrSEmA4Qrg&_hsmi=34015976
16. Cell Reports http://www.cell.com/cell-reports/home
17. Family Food Datasets: Detailed annual statistics on family food and drink purchases. https://www.gov.uk/government/statistical-data-sets/family-food-datasets
18. FAOSTAT Data: Land Use http://www.fao.org/faostat/en/#data/RL/visualize
19. Edible Insects: Future Prospects for Food and Feed Security
    http://www.fao.org/3/i3253e/i3253e00.htm
20. Talking of Taste: Umami? Kokumi? The search for new tastes goes way beyond gastronomy https://www.psychologytoday.com/ca/articles/201709/talking-taste
21. Foods of the Future: What Will We Be Eating?
    https://www.forbes.com/sites/forbesinternational/2015/11/13/foods-of-the-future-what-will-we-be-eating/#7a58ce8572c0
22. Thanksgiving Food for Thought: The Tech Helping Make Food Abundant
    https://singularityhub.com/2018/11/22/thanksgiving-food-for-thought-the-tech-helping-make-food-abundant/#sm.0001pb1slked5db111lcc8wrf59hi
23. Treating City Wastewater as a Resource
    https://www.raconteur.net/sustainability/wastewater-resource-city
24. A "Fourth-Generation" DNA Base Editor Could Replace CRISPR
    https://futurism.com/a-fourth-generation-dna-base-editor-could-replace-crispr/

25. From Pixels to Plate, Food Has Become 3D Printing's Delicious New Frontier https://www.digitaltrends.com/cool-tech/3d-food-printers-how-they-could-change-what-you-eat/
26. Future Foods: What Will We Be Eating in 20 Years' Time? http://www.bbc.com/news/magazine-18813075
27. The Future of Food https://harvardmagazine.com/2018/03/future-of-food-and-farming
28. The Future of Food – The Food of the Future http://medicalfuturist.com/the-future-of-food-the-food-of-the-future/
29. The Future of Food: Experts Predict How Our Plates Will Change http://time.com/3482452/future-of-food/
30. The Future of Food: The Stomach Wars Are o https://www.thoughtworks.com/insights/blog/future-food-stomach-wars
31. Cell Reports, Mature Human White Adipocytes Cultured under Membranes Maintain Identity, Function, and Can Transdifferentiate into Brown-like Adipocytes https://www.cell.com/cell-reports/pdf/S2211-1247(19)30342-0.pdf
32. Scientists Find Way to Convert Bad Body Fat into Good Fat https://www.aau.edu/research-scholarship/featured-research-topics/scientists-find-way-convert-bad-body-fat-good-fat
33. Genetic Engineering Could Allow Us to Treat or Even Prevent Obesity https://futurism.com/genetic-engineering-could-allow-us-to-treat-or-even-prevent-obesity/
34. Geneticists Have Used CRISPR Gene Editing to Create Crops That Grow More Food https://futurism.com/geneticists-have-used-crispr-gene-editing-to-create-crops-that-grow-more-food/
35. How Diet Influences Your Genes http://drhyman.com/blog/2017/06/23/diet-influences-genes/?utm_source=Newsletter&utm_campaign=c4a4473577-EMAIL_CAMPAIGN_2017_05_04&utm_medium=email&utm_term=0_07a277e311-c4a4473577_103149181&mc_cid=c4a4473577&mc_eid=4e77056c8f
36. How the Food You Eat Changes Your DNA with Dr. David Perlmutter https://blog.paleohacks.com/how-the-food-you-eat-changes-your-dna-with-dr-david-perlmutter/

37. Nature's Bounty: Saving the Brain with Food
https://www.psychologytoday.com/us/articles/201803/nature-s-bounty-saving-the-brain-food
38. Future Food Institute http://futurefood.network/institute/
39. It's the year 2038 – here's how we'll eat 20 years in the future
https://www.fastcompany.com/90222618/what-the-future-of-food-will-look-like-in-2038
40. Memphis Meats http://www.memphismeats.com/
41. Open Meals http://www.open-meals.com/
42. Inside the lab where Impossible Foods makes its plant-based "blood"
https://www.fastcompany.com/90264450/inside-the-lab-where-impossible-foods-makes-its-plant-based-blood
43. The ice cream of the future is here, and it has a nipple
https://www.fastcompany.com/90238527/the-ice-cream-of-the-future-is-here-and-it-has-a-nipple
44. The $700 billion case to fight food waste
https://www.fastcompany.com/90226117/food-waste-is-valued-at-1-2-trillion-a-year
45. More proof that the future of fast food is meat-free
https://www.fastcompany.com/90223180/more-proof-that-the-future-of-fast-food-is-meat-free
46. Is this the disposable cup of the future?
https://www.fastcompany.com/90207213/could-gourds-replace-plastic-cups
47. Reaping the Benefits: Science and the Sustainable Intensification of Global Agriculture https://royalsociety.org/topics-policy/publications/2009/reaping-benefits/
48. Artifacts from the Future – Tangible, Concrete, Experiential
http://www.iftf.org/foodinnovationartifacts/
49. Forecast Perspectives – Overview: Remaking Food Experiences
http://www.iftf.org/maps/resources/resources-for-seeds-of-disruption/seeds-of-disruption-forecast-perspectives-overview/

50. Seeds of Disruption: How Technology Is Remaking the Future of Food
http://www.iftf.org/maps/seeds-of-disruption/seeds-of-disruption-map/
51. Obesity and Overweight (Fact sheet)
http://www.who.int/mediacentre/factsheets/fs311/en/
52. Overweight & Obesity: Adult Obesity Facts
https://www.cdc.gov/obesity/data/adult.html
53. Robert Lustig: The man who believes sugar is poison
https://www.theguardian.com/lifeandstyle/2014/aug/24/robert-lustig-sugar-poison
54. Scientists Find Way to Convert Bad Body Fat into Good Fat: Potential for Treating Obesity Identified in Mice
https://www.eurekalert.org/pub_releases/2017-09/wuso-sfw091817.php
55. The Taste of Tomorrow: Dispatches from the Future of Food https://www.amazon.com/Taste-Tomorrow-Dispatches-Future-Food/dp/0061804215
56. Vertical Farming Is Officially Coming to Grocery Stores
https://futurism.com/vertical-farming-is-officially-coming-to-grocery-stores/
57. We May Have Uncovered the Biological Mechanism to Turn off Hunger
https://futurism.com/uncovered-biological-mechanism-turn-off-hunger/
58. We're in a New Age of Obesity. How Did It Happen? You'd Be Surprised
https://www.theguardian.com/commentisfree/2018/aug/15/age-of-obesity-shaming-overweight-people
59. Your Produce Might Soon Grow in a Warehouse Down the Block
https://futurism.com/your-produce-might-soon-grow-in-a-warehouse-down-the-block/
60. 11 things to know about global obesity
https://www.weforum.org/agenda/2016/10/11-things-to-know-about-global-obesity
61. Childhood Obesity: The Future of Children, vol. 16, no. 1, Spring 2006
https://www.brookings.edu/wpcontent/uploads/2012/04/foc_16_1_summary.pdf

62. Our Children Are Our Future http://www.healthy-holistic-living.com/childhood-obesity-statistics.html
63. Would You Like Bugs with That? Sushi Lovers More Likely to Eat Insects https://www.nzherald.co.nz/thecountry/news/article.cfm?c_id=16&objectid=12211762
64. Tackling Obesities: Future Choices https://www.gov.uk/government/collections/tackling-obesities-future-choices
65. Obesity Crisis: Future Projections 'Underestimated' http://www.bbc.co.uk/news/health-25708278
66. Obesity Trends Indicate More Than Half of US Children at Risk http://www.pharmacytimes.com/resource-centers/weight-management/obesity-trends-indicate-more-than-half-of-us-children-at-risk
67. Magic' Obesity-Eradicating Pills, Robot Doctors in Our Homes and People Living FOREVER… The Future of Healthcare Revealed https://www.thesun.co.uk/tech/4410997/magic-obesity-eradicating-pills-robot-doctors-in-our-homes-and-people-living-forever-the-future-of-healthcare-revealed/
68. Urban Obesity https://www.futureagenda.org/insight/urban-obesity
69. The Future of Pediatric Obesity https://www.primarycare.theclinics.com/article/S0095-4543(15)00087-1/pdf
70. Perspectives on Childhood Obesity Prevention: Recommendations from Public Health Research and Practice https://www.jhsph.edu/research/centers-and-institutes/johns-hopkins-center-for-a-livable-future/_pdf/research/clf_reports/childhoodobesity.pdf
71. Obesity Has Become 'A National Threat' to the UK Like Terrorism https://www.telegraph.co.uk/news/health/news/12044585/Obesity-has-become-a-national-threat-like-terrorism.htm
72. Preventing Childhood Obesity: Early-Life Messages from Epidemiology https://onlinelibrary.wiley.com/doi/full/10.1111/nbu.12277
73. Alexa, What's for Dinner Tonight? https://www.morganstanley.com/ideas/online-food-delivery-market-expands

74. Future governance options for large-scale land acquisition in Cambodia: Impacts on tree cover and tiger landscapes
https://www.sciencedirect.com/science/article/pii/S1462901118306300
75. The Land Sparing Complex: Environmental Governance, Agricultural Intensification, and State Building in the Brazilian Amazon
https://www.tandfonline.com/doi/abs/10.1080/24694452.2017.1309966?scroll=top&needAccess=true&journalCode=raag21
76. The New Climate Economy: The 2018 Report of the Global Commission on the Economy and Climate, Section Three: Food and Land Use
https://newclimateeconomy.report/2018/food-and-land-use/
77. How imperfect can land sparing be before land sharing is more favourable for wildspecies? https://besjournals.onlinelibrary.wiley.com/doi/full/10.1111/1365-2664.13282
78. Pratodomundo – Comida para 10 bilhões – Museu do Amanhã: The New Exhibition of the Museum of Tomorrow
https://pratodomundo.museudoamanha.org.br/en/
79. Food Waste Facts https://www.tristramstuart.co.uk/food-waste-facts
80. How Scientists Hacked Photosynthesis to Up Crop Yields By 40 Percent
https://singularityhub.com/2019/01/08/how-scientists-hacked-photosynthesis-to-up-crop-yields-40-percent/?utm_medium=email&utm_source=eblast&utm_campaign=fy19q1-xthinkers&utm_content=January-week-2&mkt_tok=eyJpIjoiTWpJeE56RXpOamM1T1RneSIsInQiOiJNMFwvSFh4RVhaR2FIcFg5QkNGYmNmYU9jSWZmVDNRNzJzRndoM3hkbDFnc0RhNFJOdTVKSGUwUjk0QWt1WDI1NVlZK2w3TFNuN2cyXC9wZWQ0eGo1c1dRdWNcL2tpRHhWMkJXOUEwSGcwRlwvZE84VFhyTUxlOWdVbjYyTUtnVDJEZncifQ%3D%3D#sm.0001pb1slked5db111lcc8wrf59hi
81. Food Consumption Trends and Drivers
https://www.ncbi.nlm.nih.gov/pmc/articles/PMC2935122/
82. Food Companies Are Tricking You with Words Like 'Artisanal' and 'Homemade,' Says Consumer Group https://munchies.vice.com/en_uk/article/d3kexv/food-

companies-are-tricking-you-with-words-like-artisanal-and-homemade-says-consumer-group
83. Climate Change Threatens the World's Food Supply, United Nations Warns https://www.nytimes.com/2019/08/08/climate/climate-change-food-supply.html
84. The importance of food naturalness for consumers: Results of a systematic reviewhttps://www.sciencedirect.com/science/article/pii/S092422441730122X
85. Tackling Food Waste: One of the Biggest Challenges of Our Time http://foodsustainability.eiu.com/tackling-food-waste-one-biggest-challenges-time/?utm_campaign=EP2019%20-%20Email%2017%20-%20NEW%20-%20HTML%20-%2026042019&utm_medium=email&utm_source=Eloqua
86. Growing Underground: The Hydroponic Farm Hidden 33 Metres Below London https://www.wired.co.uk/article/underground-hydroponic-farm
87. Inside London's First Underground Farm https://www.independent.co.uk/Business/indyventure/growing-underground-london-farm-food-waste-first-food-miles-a7562151.html
88. World Health Organization, Obesity and Overweight https://www.who.int/news-room/fact-sheets/detail/obesity-and-overweight
89. We're in a New Age of Obesity. How Did It Happen? You'd Be Surprised https://www.theguardian.com/commentisfree/2018/aug/15/age-of-obesity-shaming-overweight-people
90. CDC Centers for Disease Control and Prevention, Disability and Risk Factors, Overweight & Obesity https://www.cdc.gov/nchs/fastats/obesity-overweight.htm
91. CDC Centers for Disease Control and Prevention, Overweight & Obesity, Data & Statistics, Adult Obesity Fact https://www.cdc.gov/obesity/data/adult.html
92. Climate-Friendly Diets Tend to Be Good for Us https://www.futurity.org/climate-friendly-healthier-diets-1964632-2/?utm_source=Futurity+Today&utm_campaign=4ec526dce4-EMAIL_CAMPAIGN_2019_01_28_04_10&utm_medium=email&utm_term=0_e34e8ee443-4ec526dce4-206430181

93. This Part of the Brain May Shape Our Food Choices https://www.futurity.org/food-choices-brain-1892452/
94. Shoppers of the Future https://www.igd.com/research/igd-futures/shoppers-of-the-future
95. Food Fight: Why the Next Big Battle May Not Be Fought Over Treasure or Territory – But for Fish. https://foreignpolicy.com/2018/09/12/food-fight-illegal-fishing-conflict/
96. Climate Change: Report Says 'Cut Lamb and Beef' https://www.bbc.co.uk/news/amp/science-environment-46214864
97. Big Food Is Testing the Wellness Waters https://skift.com/2019/10/25/big-food-is-testing-the-wellness-waters/
98. The Future of Food - Stefan Hyttfors - Futurist Speaker https://www.youtube.com/watch?v=LiH29QAKI2c
99. The Technological Future of Food and Eating - The Medical Futurist https://www.youtube.com/watch?v=gkOZ74R6ICw
100. Futurist Thomas Frey presents the "The Future of Food and Farming" https://www.youtube.com/watch?v=lZVOwgBZNSc
101. Irwin Adam - Food Futurist https://www.youtube.com/watch?v=JnKkMZhlGk8
102. Apps step up to the plate with diet and DNA-tailored advice: New online services nudge users towards smarter shopping and eating choices https://www.ft.com/content/4577f938-2632-11e9-b20d-5376ca5216eb
103. Wellness Trends to Look Forward to in 2019 https://www.shape.com/lifestyle/mind-and-body/wellness-trends-look-forward-2019
104. Tailor-Made Nutrition is Open for (Big) Business https://www.globalwellnesssummit.com/trendium/personalized-nutrition-is-big-business/
105. Summit Trend in the News: Nutrition Gets Very Personalized https://www.globalwellnesssummit.com/trendium/technology-and-science-deliver-your-best-meal/

106. Growing Population: United Nations Department of Economic and Social Affairs, Population Division (UNDESA). 2013. *World Population Prospects: The 2012 Revision*. New York: United Nations. Total population by major area, region, and country. Medium fertility scenario.
107. Shifting Diets: Bunderson, W. T., 2012. "*Faidherbia albida*: the Malawi experience." Lilongwe, Malawi: Total LandCare
108. Food Gap: WRI analysis based on Alexandratos, N., and J. Bruinsma. 2012. *World agriculture towards 2030/2050: The 2012 revision*. Rome: FAO.
109. Food Distribution: WRI analysis based on FAO. 2012. "FAOSTAT." Rome: FAO; United Nations, Department of Economic and Social Affairs, Population Division (UNDESA). 2013. *World Population Prospects: The 2012 Revision*. New York: United Nations. Medium fertility scenario.
110. Agriculture's Environmental Footprint: WRI analysis based on IEA (2012); EIA (2012); EPA (2012); Houghton (2008); FAO (2011); FAO (2012); Foley et al. (2005).
111. Climate Change and Crop Yields: World Bank. 2010. *World Development Report 2010: Development and Climate Change*. Washington, DC: World Bank.
112. Growing Water Stress: World Resources Institute and The Coca-Cola Company. 2011. *Glob. Biogeochem. Cycles* 22: GB1003, DOI:1010.1029/2007GB002952.
113. Energy-Food Nexus: Heimlich, R., and T. Searchinger. Forthcoming. *Calculating Crop Demands for Liquid Biofuels*. Washington, DC: World Resources Institute.
114. Food and Development: World Bank. 2012.
115. Great Balancing Act: WRI.
116. Annual Crop Production: WRI analysis based on Bruinsma, J. 2009. *The Resource Outlook to 2050: By how much do land, water, and crop yields need to increase by 2050?* Rome: FAO; Alexandratos, N., and J. Bruinsma. 2012. *World agriculture towards 2030/2050: The 2012 revision*. Rome: FAO.
117. GHG Emissions from Animal Products: GLEAM in Gerber, P. J., H. Steinfeld, B. Henderson, A. Mottet, C. Opio, J. Dijkman, A. Falcucci, and G. Tempio. 2013. *Tackling climate change through livestock: A global assessment of emissions and mitigation opportunities*. Rome: FAO.

118. Current and Projected Fertility Rates: United Nations Department of Economic and Social Affairs, Population Division (UNDESA). 2013. *World Population Prospects: The 2012 Revision*. New York: United Nations. Total fertility by major area, region, and country. Medium fertility scenario.
119. Cereal Yields: Derived from FAO. 2012. "FAOSTAT." Rome: FAO.
120. Maize Yields in Malawi: Bunderson, W. T. 2012. "*Faidherbia albida*: the Malawi experience." Lilongwe, Malawi: Total LandCare.
121. Degraded Lands in Kalimantan: Gingold, B. et al. 2012. *How to Identify Degraded Land for Sustainable Palm Oil in Indonesia*. Washington, DC: World Resources Institute.
122. Fish Production: FAO. 2012. "FishStatJ." Rome: FAO.
123. Closing the Food Gap: WRI analysis based on Alexandratos, N., and J. Bruinsma. 2012. *World agriculture towards 2030/2050: The 2012 revision*. Rome: FAO.

www.ingramcontent.com/pod-product-compliance
Lightning Source LLC
Chambersburg PA
CBHW040314220526
45473CB00009B/2429